Blockchain Technology and Applications

Blockchain Technology and Applications

Edited by

Pethuru Raj
Kavita Saini
Chellammal Surianarayanan

CRC Press is an imprint of the
Taylor & Francis Group, an informa business
AN AUERBACH BOOK

First Edition published 2021
by CRC Press
6000 Broken Sound Parkway NW, Suite 300, Boca Raton, FL 33487-2742
and by CRC Press
2 Park Square, Milton Park, Abingdon, Oxon, OX14 4RN

ISBN: 978-0-367-53340-3 (hbk)
ISBN: 978-0-367-54275-7 (pbk)
ISBN: 978-1-003-08148-7 (ebk)

Visit the [companion website/eResources]: [insert comp website/eResources URL]

Contents

Editors

Pethuru Raj works as the chief architect in the Site Reliability Engineering (SRE) division of Reliance Jio Infocomm Ltd. (RJIL), Bangalore. He previously worked as a cloud infrastructure architect in the IBM Global Cloud Center of Excellence (CoE), IBM India Bangalore for four years. Prior to that, he had a long stint as a TOGAF-certified enterprise architecture (EA) consultant in Wipro Consulting Services (WCS) Division. He also worked as a lead architect in the corporate research (CR) division of Robert Bosch, Bangalore. In total, he has gained more than 17 years of IT industry experience and 8 years of research experience.

He finished the CSIR-sponsored PhD degree at Anna University, Chennai, and continued with the UGC-sponsored postdoctoral research in the Department of Computer Science and Automation, Indian Institute of Science, Bangalore. Thereafter, he was granted a couple of international research fellowships (JSPS and JST) to work as a research scientist for 3.5 years in two leading Japanese universities. He has published more than 30 research papers in peer-reviewed journals such as IEEE, ACM, Springer-Verlag, Inderscience, etc. He has contributed 35 book chapters thus far for various technology books edited by highly acclaimed and accomplished professors and professionals.

Kavita Saini is presently working as an associate professor, School of Computing Science and Engineering, Galgotias University, India. She received a PhD degree from Banasthali Vidyapeeth, Banasthali. She has 16 years of teaching and research experience and supervising M.Tech. and PhD scholars in various areas.

Her research interests include web-based instructional systems (WBIs), software engineering, blockchain, and database. She has published various books for UG and PG courses for various universities including M.D. University, Rohtak, and Punjab Technical University, Jallandhar.

She has published 22 research papers in national and international journals and conferences and delivered a technical talk on "Blockchain: An Emerging Technology," "Web to Deep Web," and other emerging areas.

Chellammal Surianarayanan is an assistant professor of Computer Science in Bharathidasan University Constituent Arts & Science College, Tiruchirappalli, Tamil Nadu, India. She earned Master of Science in Physics, Master of Technology in Information Technology, and a doctorate in computer science. She worked on the discovery and selection of semantic web services. She has published research papers in Springer *Service-Oriented Computing and Applications*, *IEEE Transactions on Services Computing*, *International Journal of Computational Science*, *Inderscience* and *SCIT Journal* (Symbiosis Centre for Information Technology), etc. She has produced book chapters with IGI Global and CRC Press. She is a life member in professional bodies such as the Computer Society of India, IAENG, etc.

Before coming to academic service, Chellammal Surianarayanan served as Scientific Officer in the Indira Gandhi Centre for Atomic Research, Department of Atomic Energy, Government of India, Kalpakkam, Tamil Nadu, India. She was involved in the research and development of various need-based embedded systems and software applications. Her remarkable contributions include the development of an embedded system for a lead shield integrity assessment system, portable automatic air sampling equipment, an embedded system of detection of lymphatic filariasis in its early stage, and the development of data-logging software applications for atmospheric dispersion studies. In total, she has 21+ years of academic and industrial experience

Contributors

M. Vivek Anand is a research scholar in the Department of CSE, Galgotias University, Greater Noida, Uttar Pradesh, India. He received an M.E. in Software Engineering from Anna University, Chennai, Tamil Nadu, in 2013, and a Bachelor of Engineering in the stream of Computer Science from Anna University, Coimbatore, Tamil Nadu, in 2011. He has more than five years of teaching experience. His research interests are the Internet of Things and blockchain.

K. P. Arjun is a research scholar in the Department of CSE, Galgotias University, Greater Noida, Uttar Pradesh, India. He received an M.Tech. in Computer Science and Engineering from the University of Calicut, Kerala, in 2016. His research interests are big data analytics, cloud computing, artificial intelligence, machine learning and deep learning. He has authored over five research papers in various national and international journals and conferences. His publications are indexed in SCI, Scopus, Web of Science, DBLP and Google Scholar.

D. Peter Augustine is presently working as associate professor, Department of Computer Science, at CHRIST (Deemed to be University). He received a B.Sc. degree from St. Xavier's College, Palayamkottai, in 1997, an M.C.A. degree from Manonmaniam Sundaranar University, Tirunelveli, in 2000, and completed his doctorate degree at CHRIST (Deemed to be University), Bangalore. He has overall experience of 19 years with 3 years in industry, and 16 years in the teaching field. His research interests include cloud computing, medical image processing and big data analytics. He has published papers in Scopus-indexed international journals and conferences.

His research interests include artificial intelligence, IoT and big data analysis in the area of healthcare and data mining, and human computer interaction. He has written chapters for books focusing on some of the emerging technologies such as IoT, data analytics and science, blockchain and digital twin.

B. Balamurugan completed his PhD at VIT University, Vellore, and is currently working as a Professor in Galgotias University, Greater Noida, Uttar Pradesh. He has 15 years of teaching experience in the field of computer science. His area of interest lies in the fields of the Internet of Things, big data and networking. He has published more than 100 international journal papers and contributed book chapters.

Broto Rauth Bhardwaj works as Head, Research and Entrepreneurship Development at Bharati Vidyapeeth University, New Delhi. She completed her post-doctoral thesis at University of California, Los Angeles (UCLA), USA. She has a PhD and MBA from IIT, Delhi. A gold medalist from IIT Delhi, she has more than 16 years of industry and teaching experience. She has published more than 100 papers in national and international journals. She has eight PhD scholars under her. One of them has been awarded and two have submitted their theses. She has organized management development programs, faculty development programs and motivational programs for the US government and has undertaken such assignments with Hopkins County College, USA.

Rahul Chauhan is currently a student and pursuing a B.Tech. in Computer Science from Galgotias University, Uttar Pradesh, India. He is also pursuing a specialization in Data Analytics. He is currently in the team of bdverse, where he is developing a package to analyze biodiversity data. His area of interest is data science and making voice-based applications. He was selected in Google Summer of Code 2019 and is currently an Alexa Student Influencer. He loves teaching and has conducted hundreds of workshops all over India. He is currently working on creating a shiny dashboard that allows a non-technical person to use the capabilities of R in visualizing datasets.

Saugata Dutta received his Bachelor of Computer Science from Alagappa University, Tamil Nadu, India, and received his Master of Computer Science from Punjab Technical University, Punjab, India. He has four publications in the field of network security and blockchain technology. He is presently doing research in blockchain technology at Galgotias University, Greater Noida, India, and is associated with an IT company as Head of Department in Information Technology.

N. S. Gowri Ganesh received his B.E. degree in Electronics and Communication Engineering and M.E. in Computer Science and Engineering from Bharathiar University, Coimbatore, India, in 1993 and 2000, respectively, as well as a PhD in Computer Science and Engineering from Anna University, Chennai, in 2015.

In 1993, he joined the Numeric Power Systems Ltd., an uninterrupted power supply manufacturing industry and worked in the Research and Development, Production and Quality Department. He worked as Senior Technical support engineer for technical support (offshore) cases for Sybase Inc. He worked as a Lecturer at the Sathyabama Institute of Science and Technology, Chennai, India, and R.R. Engineering College, Tiruchengode, India, between 2001 and 2004. In 2004, he worked as Senior Lecturer at Siva Subramania Nadar College, Chennai, India. He joined the Centre for Development of Advanced Computing (C-DAC) as Senior Engineer in 2006. He worked in various open source projects and was involved with the first seed version of the Indian version of Linux – BOSS (www.bosslinux.in). He was the SEPG Head for acquiring CMMI Level 3 for C-DAC, Chennai, India. He is presently working as Professor and Head of Department of Information Technology, at Malla Reddy College of Engineering and Technology, Hyderabad, India, from 2016. He is a Life Member of the Indian Society for Technical Education (ISTE). He is also a member of the International Association of Engineers (IAENG) and Computer Science Teachers association (CSTeachers.org). He was one of the coordinators for the First International Conference on Soft Computing and Signal Processing (ICSCSP 2018) that Springer held at the Malla Reddy College of Engineering and Technology, Hyderabad. His research interests are cloud computing, Provenance, Blockchain, IoT, and web services.

R. Indrakumari is presently working as assistant professor, Department of Computer Science and Engineering, at Galgotias University, Uttar Pradesh, India. She received a B.E. degree from Madurai Kamaraj University, India, in 2001 and an M.Tech. in Computer Science and Information Technology from Manonmaniam Sundaranar University, Tirunelveli. She has overall experience of 15 years, with 4 years in industry and 11 years in the teaching field. Her research interests include data mining, big data, data warehousing and tools like Tableau and QlikView. She has published papers in international journals and conferences.

Pramod Mathew Jacob completed his B.Tech. in Computer Science and Engineering from Kerala University, Thiruvananthapuram, India. He possesses an M.Tech. in Software Engineering from SRM Institute of Science and Technology, Chennai, India. Presently he is working as assistant professor in Providence College of Engineering, Chengannur, Kerala, India. He is also pursuing a PhD at Vellore Institute of Technology (Deemed to be University), Vellore, India. He has teaching experience of six years and research experience of three years. He has published ten papers in various international journals and conferences. His areas of interest include software engineering, software testing and the Internet of Things.

Vishal Jain is currently working as associate professor with Bharati Vidyapeeth's Institute of Computer Applications and Management (BVICAM), New Delhi, Affiliated to GGSIPU and Accredited by AICTE, from July, 2017 to date. He joined BVICAM, New Delhi, in 2020 and worked as Assistant Professor from August, 2010, to July, 2017. He received the Young Active Member award for the year 2012/2013 from the Computer Society of India. His research areas include the semantic web, ontology engineering, cloud computing, big data analytics and ad-hoc networks.

Rajesh Kaluri completed his PhD in Computer Vision at VIT University, India. He obtained a B.Tech. in CSE from JNTU, Hyderabad, India, and did an M.Tech. in CSE from ANU, Guntur, India. Currently he is working as an assistant professor (senior) in the School of Information Technology and Engineering, VIT University, India. He has 8.5 years of teaching experience. He was a visiting professor in Guangdong University of Technology, China, in 2015 and 2016. His current research is in the areas of computer vision and human–computer interaction. He has published research papers in various reputed international journals.

K. Sampath Kumar is a professor in the School of Computing Science and Engineering, Galgotias University, Greater Noida, UP, NCR Delhi, India. He completed his PhD in Data Mining from Anna University, Chennai, India, and obtained his M.E. from Sathyabama University, Chennai, India, in 2005. He has over 17 years of teaching experience. His expertise is in big data, cloud computing, IoT, artificial intelligence and real-time systems. He has published more than 50 research articles in international journals and conferences.

Prasanna Mani completed his M.S. in Computer Science and Engineering and received his doctorate in Software Engineering, both from Anna University, Chennai, India. Presently he is working as Associate Professor in Vellore Institute of Technology (Deemed to be University), Vellore, Tamil Nadu, India. He possesses teaching experience of about 20 years from various reputed colleges and universities. He has published nearly 25 papers in various national and international journals. He is guiding research scholars in the area of software testing and is an eminent reviewer of various international journals. He has also authored a book for cracking interview questions of C programming. His areas of interest include software engineering, software testing, the Internet of Things, etc.

M. R. Manu is currently working as a Computer Science Teacher in the Ministry of Education, Abudabi, UAE. He worked as an assistant professor in the School of Computing Science and Engineering, Galgotias University, NCR Delhi, India. He completed an M.E. in Computer Science and Engineering from Anna University Taramani Campus, Tamil Nadu, India, and is currently pursuing a PhD in Computer Science and Engineering from Galgotias University, NCR Delhi, India. His areas of interest are big data, networks and network security. He has undertaken different research projects in networks specialization and published 16 papers in various international and national journals. He is currently writing a monograph and book chapters with CRC Press, Springer, and Elsevier publishers.

Namya Musthafa is currently pursuing an M.E. in Computer Science and Engineering from the Royal College of Engineering and Technology, Kerala, India, where she received a B.Tech. in Computer Science and Engineering in 2017. Her areas of interest are big data and machine learning. She has published in around three international journals indexed by Scopus.

C. Navaneethan is currently working as associate professor in the School of Information Technology and Engineering, Vellore Institute of Technology, Vellore, Tamil Nadu, India. He pursued his undergraduate degree in Engineering with Computer Science and Engineering as Specialization in April 2004. He was awarded Honors in his M.E. CSE in July 2006 and completed his PhD in wireless sensor networks from Anna University, Chennai, India, in 2017. He has published and presented in many national and international journals/conferences. His current areas of research activities include wireless sensor networks and network security. He is a research paper reviewer in conferences at the national and international levels and also a life member in professional bodies like ISCA, IAENG, IACSIT and CSTA.

Pooja Saigal is presently working as associate professor, School of Information Technology, at Vivekananda Institute of Professional Studies (affiliated to GGSIP University), New Delhi, India. She received her PhD degree in Machine Learning from South Asian University (established by SAARC), New Delhi, India, in 2018. She holds a Master's degree (2004) and Bachelor's degree (2001) in Computer Applications. She is University Topper of M.C.A. class of 2004 and was awarded a Gold Medal by the erstwhile President of India, Dr. APJ Abdul Kalam, for her outstanding performance in the M.C.A. program. She got Distinction in all 32 courses of her M.C.A. She secured First Rank in University in her B.C.A. and was awarded by the Chief Minister of Haryana. She has total experience of over 16 years in teaching information technology and computer science subjects at postgraduate and undergraduate levels, with successful records of accomplishments. This includes research experience of four years at South Asian University, New Delhi, India, and industry experience of six months with the National Informatics Center (NIC), Delhi Secretariat, New Delhi. Her research interests include artificial intelligence, machine learning, optimization and image processing. She is UGC-NET qualified in computer science. She is a reviewer of SCI-indexed journals: *Neurocomputing* (Elsevier), *Neural Networks* (Elsevier) and *IEEE Transactions on Cybernetics*, among others. She has published research papers in international journals and conferences of high repute.

T. Poongodi works as an associate professor in the School of Computing Science and Engineering, Galgotias University, NCR Delhi, India. She completed her PhD in Information Technology (Information and Communication Engineering) from Anna University, Tamil Nadu, India. Her main research areas are big data, the Internet of Things, ad-hoc networks, network security and cloud computing. She is a pioneer researcher in the areas of big data, wireless networks and the Internet of Things and has published more than 25 papers in various international journals. She has presented papers at national/international conferences, published book chapters in CRC Press, IGI Global and Springer, and edited books.

S. Prasanna received his B.E. degree in Computer Science and Engineering from University of Madras, Chennai, India, in 2001, and an M.E. degree in Computer Science and Engineering from Anna University, Chennai, India, in 2006. He completed his Doctor of Philosophy in Vellore Institute of Technology, Vellore, India. He has published more than 15 papers in reputed journals and conferences. He is currently associate professor with the School of Information Technology and Engineering in Vellore Institute of Technology. His areas of interest are data mining, soft computing, artificial intelligence and blockchain technology.

N. M. Sreenarayanan is a research scholar in the Department of CSE, Galgotias University, Greater Noida, Uttar Pradesh, India. He received an M.Tech. in Computer Science and Engineering from the University of Calicut, Kerala, India, 2016. His research interests are artificial intelligence, machine learning, neural networks and deep learning. He has authored over five research papers in various national and international journals and conferences. His publications are indexed in SCI, Scopus, Web of Science, DBLP and Google Scholar.

T. Subha is presently working as associate professor, Department of Information Technology, at Sri Sai Ram Engineering College/Anna University, Chennai, India. She received a B.E. in Computer Science and Engineering from Bharathidasan University, Tiruchirappalli, India, in 2000, an M.Tech. in Information Technology from Sathyabama University, Chennai, India, in 2009, and completed her doctorate degree in Anna University, 2018. She has overall experience of 19 years in teaching. Her research interests include network security, cloud computing and cyber security. She has published 40 papers in international journals and conferences.

R. Sujatha completed a PhD at Vellore Institute of Technology, Vellore, India, in 2017 in the area of data mining. She received her M.E. degree in computer science from Anna University, Chennai, India, in 2009 with university ninth rank and completed a Master of Financial Management from Pondicherry University, India, in 2005. She received her B.E. degree in computer science from Madras University, Chennai, India, in 2001. She has 15 years of teaching experience and serves as an associate professor in the School of Information Technology and Engineering in Vellore Institute of Technology, Vellore, India. She has organized and attended a number of workshops and faculty development programs. She is actively involved in the growth of the institute by contributing in various committees at both the academic and administrative levels. She gives technical talks in colleges for symposiums and various sessions. She acts as an advisory, editorial member and technical committee member in conferences conducted in other educational institutions and in-house, too. She has published a book titled *Software Project Management* for college students. She has also published research articles and papers in reputed journals. She used to guide projects for undergraduate and postgraduate students. She currently guides doctoral students. She is interested in exploring different places and visit the same to know about the culture and people of various areas. Her areas of research include data mining, machine learning, image processing and management of information systems.

R. Viswanathan works as a professor in the School of Computing Science and Engineering, Galgotias University, Greater Noida, Uttar Pradesh, India. He received his doctoral degree in Computer Science and Engineering from VIT University in 2016. He has 10+ years of experience working in academia, research, teaching and academic administration. His current research interests include artificial intelligence, machine learning, cloud computing, Internet of Things and data mining. He guided more than five research scholars for PhD in various domains. He has authored over 50 research papers in various national and international journals and conferences.

Ritika Wason works as associate professor at Bharati Vidyapeeth's Institute of Information Technology and Management, New Delhi, India. She has also been the recipient of many awards and honors for her academic and research work during her academic career. A certified Mendeley (a reference management tool) trainer, she has successfully conducted many workshops offering Mendeley training to participants. She also serves several other responsibilities, for example as resident editorial board member for the *International Journal of Information Technology* [an official publication of Bharati Vidyapeeth's Institute of Computer Applications and Management (BVICAM)], New Delhi, by Springer (the world's largest publisher of scientific documents) w.e.f. Volume 09 Issue 01 from January, 2017 under ISSN: 2511-2104 (Print Version); ISSN: 2511-2112 (Electronic Version). She is also the editor for the monthly magazine *CSI Communications* of the esteemed Computer Society of India (CSI). With almost a decade of teaching experience, she has always been an active researcher. She has authored four books on Software Testing and .Net Technology. She has also published many research papers as well as book chapters in national and international journals, conference proceedings, bulletins and edited books. A life member of the Computer Society of India (CSI) and Indian Society for Technical Education (ISTE), she also serves as a reviewer of international/national journals and a member of the technical committee of several international/national conferences. Since June, 2019, she has also been working as the editor for CSI Communications, a monthly theme-based national publication covering technical articles of current interests and reports of conferences, symposia and seminars.

Chapter 1

Distributed Computing and/or Distributed Database Systems

K. P. Arjun, N. M. Sreenarayanan,
K. Sampath Kumar, and R. Viswanathan

Contents

1.1 Introduction to Computing

Computing involves process-oriented step-by-step tasks to complete a goal-oriented computation. A goal is not a simple or single rather there may be more than one goal. Normally we can say that a goal is a complex operation that is processed using a computer. A normal computer contains hardware and software; and computing can also involve more than one computing environment in hardware like workstations, servers, clients and other intermediate nodes and software like a workstation Operating System, server operating system and other computing software. The computing in our daily life includes sending emails, playing games or making phone calls; these are different kinds of computing examples at different contextual levels. Depending on the processing speed and size, computers are categorized into different types like supercomputers, mainframes, minicomputers and microcomputers. The computing power of a device is directly proportional to its data-storing capacity.

All software is developed in a sequential way which means that before developing software to solve a large problem, we split the problem into smaller sub problems. These sub problems broken down step by step or in a flowchart are called algorithms. These algorithms are executed by the central processing unit (CPU). We can call this serial computing, as the main task is divided into a number of small instructions, then these instructions are executed one by one. But in the main, this serial communication is a huge waste of the hardware other than the CPU. The CPU is continuously taking instructions and processing those instructions. The hardware contributing to processing that specific hardware is used for that particular time only, and for the remaining time that hardware is idle.

So to overcome the deficiencies in resource utilization and improve the computing power we moved into another era of computing called parallel computing and distributed computing. The insight of distributed computing is in solving more complex and larger computational problems with the help of more than one computational system. The computational problem is divided into many tasks, each of which is executed in different computational systems that are located in different regions.

1.2 Evolution of Distributed Computing

Distributed computing [1] is concurrent processing of multiple processes at the same time. Distributed computing works on various very important concepts like multiprogramming and multitask programming. Finally distributed computing has been included in the branch of computer science and engineering since the 1970s. Since then many international conferences like the Symposium on Principles of Distributed Computing (PODC), International Symposium on Distributed Computing (DISC), etc. and international workshops like the International Workshop on Distributed Algorithms [2] on Graphs have been conducted.

1.2.1 Centralized Computing

The name "centralized computing" refers to computing that occurs in a central situated machine. The specifications of the central computing server machine include high computing capabilities and sophisticated software. All other computers are attached to the central situated machine and communicate through terminals. The centralized machine [3] itself controls and manages the peripherals, some of which are physically connected and some of which are attached via terminals.

The main advantage of centralized system is greater security compared to other types of computing because the processing is only done at the centrally located machine. All the connected machines can access the centralized processing machine and start processing their own task by using terminals. If one terminal goes down, then the user can use another terminal and log in again. All the user-related files are still available with that particular user login. The user can resume their session and complete the task.

The main and most important disadvantage of the centralized computing system is that all computing and storage is done at centrally located machine. If the machine fails or crashes the entire system will go down. It affects the performance evaluation on unavailability of service.

Figure 1.1 shows a block diagram of centralized computing. Centralized systems are somewhat related to client–server programming [5]. The client has minimum computing power, but for advanced and high-level computing, client requests for the server. The server computes the request received from the client and sends the response back to the client.

1.2.2 Decentralized Computing

In centralized computing, a centrally located powerful system provides computing services to all other nodes connected. The disadvantage is that all processing power is located at one entity. Alternatively, the burden at the central level can be shared by the nodes connected on the network. In decentralized computing

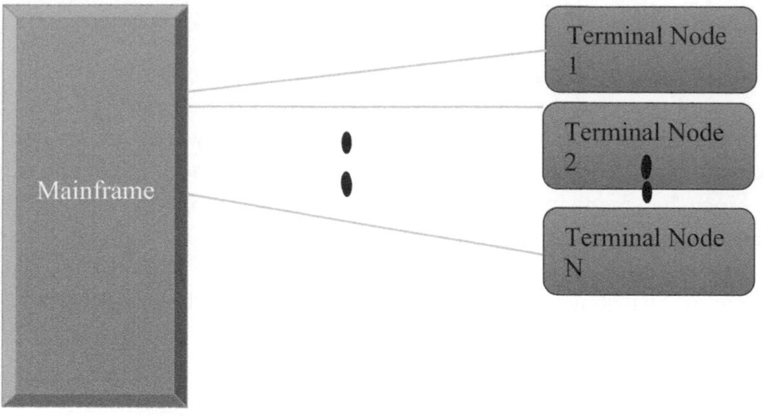

Figure 1.1 Centralized computing.

[6] a single server is not responsible for the whole task. The whole workload is distributed to the computing nodes so that each computing node has equal processing power.

1.2.3 Parallel Computing

To overcoming the deficiencies in resource utilization and improve the computing power we moved into another era of computing called parallel computing. The name "parallel" means that more than one instruction can be executed simultaneously. It requires the configuration of a number of computing engines (normally called "processors") and related hardware and also software configuration.

1.3 High-Performance Distributed and Parallel Computing

1.3.1 Parallel Computing

In a CPU, a main task is divided into a number of small instructions, and then these instructions are executed one by one. The main problem with the serial communication is wastage of large amount of resources in terms of hardware and software resources. CPU continuously receives instructions and process them. The hardware involved in serial processing remains idle in case there are no instructions to be processed.

To overcome the deficiencies in resource utilization and to improve the computing power we moved into another era of computing called parallel computing [7]. The name "parallel" means that more than one instruction can be executed simultaneously. It requires the configuration of a number of computing engines

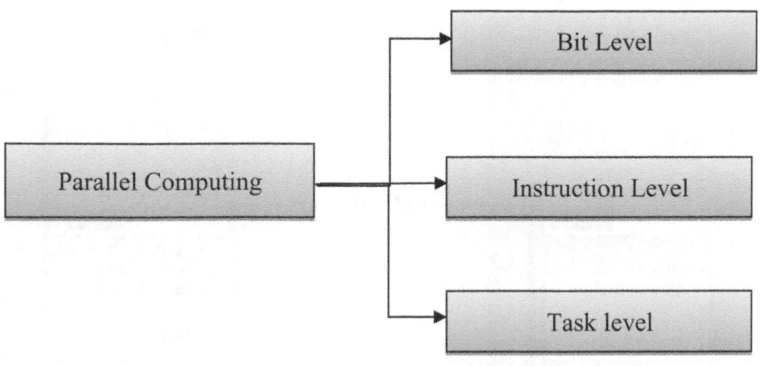

Figure 1.2 Different levels of parallel computing.

(normally called "processors") and related hardware and also software configuration. Figure 1.2 shows the levels of parallel computing.

1.3.1.1 Bit-Level, Instruction-Level and Task-Level Parallel Computing

Figure 1.1 depicts the parallel computing at various levels. The levels are bit-level, instruction-level and task-level. It is a complex type of computing because here we are adding more than one processor and the processors are supporting hardware and software. So in serial we deal with only one instruction and processor, but the challenges split the whole work into small pieces, and these small tasks are given to different computational machines. Every computation machine is independent and concurrently processing with the help of the others. Each machine deals with its own task and finally collaborates with the others as a single unit. Parallel computing added over all coordination of the execution engines [8] is one of the multifaceted problems. Parallel computing can be utilized to convert real-world scenarios to more convenient formats.

The main utility of parallel computing is in solving real-world problem-as more complex, independent and unrelated events will occur at the same time, for example, galaxy formation, planetary movements, climate changes, road traffic, weather, etc.

The advantage of fast computing is helpful in various high-end applications, for example, faster networks, high speed data transfer, distributed systems and multiprocessor computing [10], etc.

1.3.2 Distributed Computing

The distributed computing insight lies in solving more complex and larger computational problems with the help of more than one computational system. The computational problem is divided into many tasks, each of which is executed in

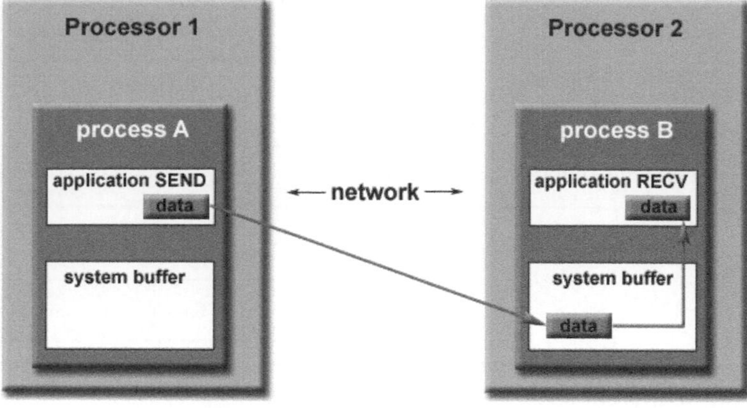

Figure 1.3 Message-passing method.

different computational systems that are located in different regions. Different computational systems located at different places communicate through strong base network communication technology. There are many communication mechanisms that have been adopted for strong and secure communications like message passing, RPC and HTTP mechanisms, etc.

Another way we can describe distributed computing is as different computational engines which are all autonomous, physically present in different geographical areas, and communicating with the help of a computer network. Each computational engine is called an autonomous system. Each autonomous system has its own hardware and software. Actually they will not share their hardware or software with another system that is located in another region. But they are continuously communicating by using the message-passing mechanism.

The main idea behind distributed computing is overcoming the limitations of computing like low processing power, speed and memory. Each computer is connected by using a single network. The duties of each computing engine are to do the assigned jobs and communicate to peer computers that are connected in the network.

A feature of the connected nodes or computers is that each one has its own hardware including memory, processor and IO devices, and software like operating systems and distributed software.

The entirety of communication happens through the message-passing mechanism. Figure 1.3 represents the message-passing mechanism.

1.3.3 Architecture of Distributed Computing

1.3.3.1 Physical Architecture of Distributed Computing

Distributed computing has much architecture related to the application and complexity of algorithms proposed at the software and hardware levels. At the high

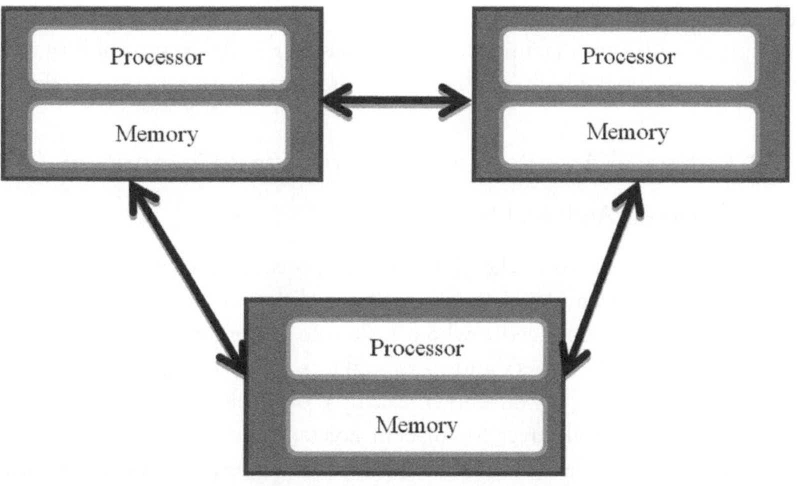

Figure 1.4 Physical block diagram of distributed computing.

model, interconnection of running state of the processes on the connected CPUs in the network. Figure 1.4 shows a physical representation of distributed computing.

All distributed computing uses one of the architecture types listed below. Each architecture type is slightly different from the others depending on the basis of the computing.

There are mainly two type of architecture; we can normally say that one is tightly coupled or loosely coupled. The name "tightly coupled distributed architecture" means that all nodes or machines are connected through a highly integrated network. It seems like all the computing engines work as a single machine. This architecture creates an illusion of a single machine but in the background different machines are connected via a fast network and memory is shared through distributed shared memory (DSM) without using the message-passing technology. Distributed shared memory (DSM) creates an illusion in this architecture of sharing memory in a network of connected nodes. Actually sharing memory is a big challenge because we have to consider the traffic across the network. The next architecture, "loosely coupled", does not share any hardware like memory processing power. The nodes just communicate together.

Other variations of architecture are client–server, three-tier, n-tier and peer-to-peer. The first one, client–server architecture, involves normal communication between the client and server. The client requests data from the server and then formats and displays it to the user. The second type is normally used for web application development. The result of this architecture simplifies the web application development. The third type is n-tier architecture, which is also used for enterprise web application development. This type of architecture is highly responsible for the success of the software framework for creating web applications. The last type, peer-to-peer architecture, includes any specific system or one system that provides services or manages network resources. All the work is equally divided among all the machines and each machine will serve that particular responsibility assigned only for that machine, which is called a peer. It acts as both server and client.

1.3.3.2 Software Architecture of Distributed Computing

1.3.3.2.1 Layered Architectures

Layered architectures involve the division of responsibilities among software components and the placement of components at different locations in computers. Layered architectures divide the whole task into different levels, and each level communicates with the others and gives services to both upper layers and lower layers. The OSI model is a well-known example of layered architecture. The communication between each layer to adjacent layers, either the upper layer or lower layer, is in sequential order. So the communication request follows in the bottom to top order, and the response follows from top to bottom order. Figure 1.5 shows the layered architecture of distributed computing.

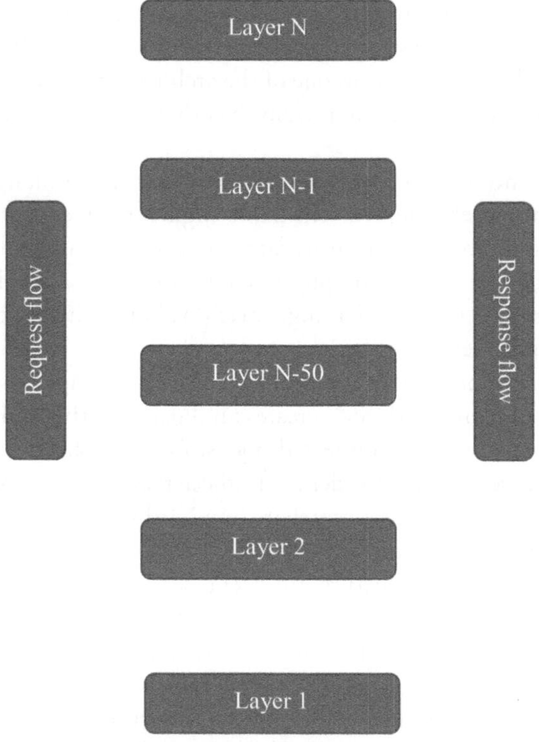

Figure 1.5 Layered architecture of distributed computing.

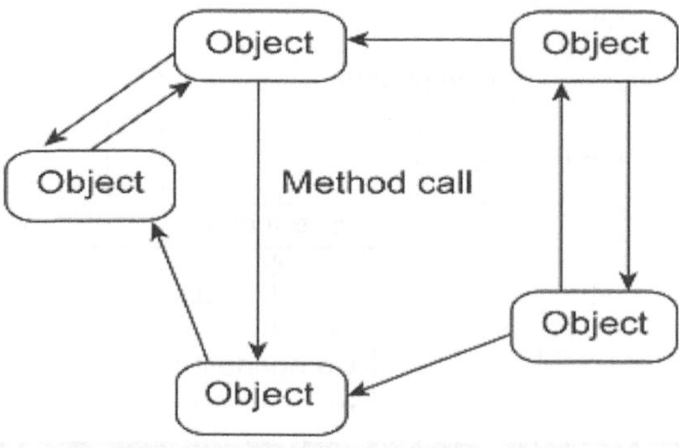

Figure 1.6 Object-based architecture of distributed computing.

The advantage of this architecture is that there is a sequential order for request and response. Each layer has its already predefined duty so there is no confusion about processing the request. We can easily update or replace each layer according to our application without affecting the entire architecture. Figure 1.6 represents the basic architecture style of a distributed system.

1.3.3.2.2 Object-Based Architectures

This style of architecture is used for loosely coupled arrangements of systems. This loosely coupled system [12] cannot follow the sequential order architecture like layered. In this architecture each component is referred to as an object; each object in a system communicates with other objects through an interface.

Objects are the incorporation of data and methods into a single unit. Communication flows from one object of a system A to an object in a system B through remote procedure call. Examples of this method are CORBA, DCOM, .Net Remoting and Java RMI. It is the one of most important architecture types in large software systems.

1.3.3.2.3 Event-Based Architectures

Nodes or components communicate on the basis of the proliferation of events. Components are connected through an event bus. An event bus carries the published and subscribed events from other components. The main advantage of this architecture is decoupled space. There is no need for the communicating components to explicitly refer to each other. Another important aspect is that it is coupled in time which means components can communicate at the same time. Figure 1.7 represents event-based architecture of distributed computing.

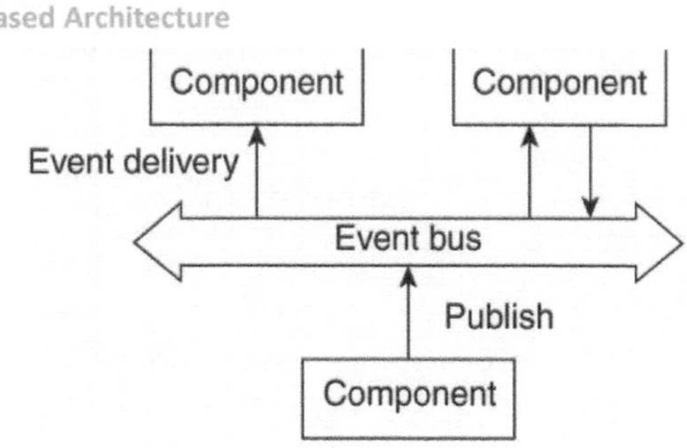

Figure 1.7 Event-based architecture of distributed computing.

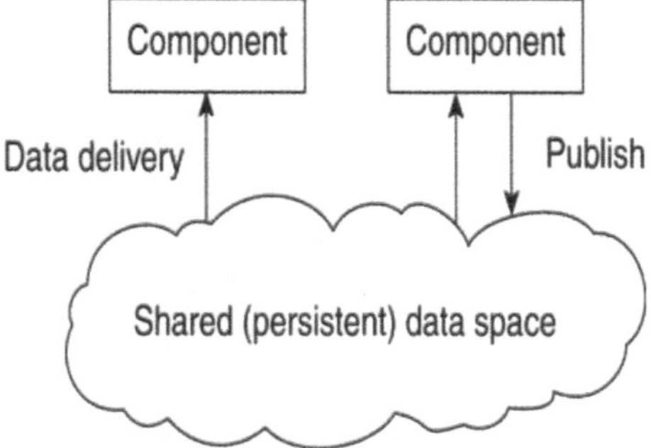

Figure 1.8 Shared data space architecture of distributed computing.

1.3.3.2.4 Shared Data Space Architectures

This is also called data-centered architecture. Here a common repository is shared between all the components that are connected in the network. This common repository has two states, either active or passive. The repository is like a database. Information from all nodes is persistently stored. The shared repository contains persistent data. The main idea is that subscribed components can send and receive data accordingly. Figure 1.8 represents shared data space architecture of distributed computing.

1.4 Comparison of Distributed Computing with State of the Art

1.4.1 Distributed Computing versus Parallel Computing

The results of distributed computing and parallel computing are same in the aspect of efficiency ant performance because both of them are interrelated apart from differences in the the placement of the hardware. In a distributed system, the computers are placed in different locations and communicate through a network. But in parallel computing all the computing hardware is combined to make a single device. In parallel computing a huge single memory is shared between computing engines, i.e. processors. Each computing station utilizes that memory with efficient synchronization. Here each processor works independently of the others. In distributed computing, each computing node has its own processor and memory like single autonomous computing nodes. The advantages of parallel and distributed computing are high-performance parallel computation [13] by using shared-memory multiprocessors and the use of parallel computing algorithms, while the coordination of a large-scale distributed system uses distributed algorithms.

1.4.2 Distributed Database Systems

In the above we discussed distributed computing and its features. In all computing methods the data are stored in a centralized fashion and computing is done in a distributed fashion. A distributed database management systems (DDBMS) is a collaboration of multiple databases that are located in different physical locations and connected through a network. These distributed databases are locally interlinked or part of a whole database system. The distributed database systems are widely used in data warehousing. Distributed databases are mainly used to manage data in networks, data confidentiality and data integrity.

1.4.3 Traditional versus Distributed Databases

Database systems are the collection of data, storage of data, management of data and finally distribution of the data to various related applications. In the past, punch cards were used for data storage. The first database was designed by Charles W. Bachman in 1960. Next the well-known company IBM implemented their own DBMS called IMS. Likewise many other companies released their own paid and unpaid software in the market, and different types of DBMS are also available.

The difference between the traditional DBMS and distributed DBMS is that distributed DBMS are the modified or latest updated version of traditional DBMS. In each development of DBMS, introduced new features were introduced that were very useful for the users as well. Nowadays many database products are available in the market. The main difference is that the traditional database management systems used only a single machine and a single software instant can access the database. These problems are addressed by distributed computing, as databases are available in different machines connected through a network. Any device can access the distributed database [14] within the network software. All types of queries can be generated from different machines connected in the network, and the distributed database system can execute the query and return back the result.

1.4.4 Distributed Computing and Blockchain

Distributed computing methods are one of the basic computing principles that drive the blockchain mechanism. Generally now everyone has a basic idea of a blockchain as a large network of computers which can authenticate and verify huge transactions. However, the internal mechanism of distributed computing can lead to a better base for blockchain technology. It can also help to make more information by focusing on working scenarios of distributed computing technologies.

1.5 Distributed Computing Environment of Blockchain

Generally, distributed computing methods are like a network of computers working together as a single system. The systems can be located close to one another and with a wired network as part of a single local network. Other networks such as blockchains widely use geographically dispersed computers networks.

Distributed computing has been used for far longer than blockchain mechanisms. The use of computers in education and research grew very early, requiring computers to connect to one another, sharing hardware such as memory and printers. In the 1970s the first local area networks were established with many systems. The first distributed computing machines were local area networks such as Ethernet, a group of networking hardware technologies developed by Xerox. Now it is widely distributed and everyone make use of it. Each time you join a new Wi-Fi connection, you are entering into a new computer network scenario.

In 21st century, the usage of distributed systems and distributed computing technologies has vital roles in solving real-world problems. Each unit of problem phases is connected to the others and finally derive appropriate solutions (Figure 1.9).

A blockchain is a peer-to-peer (P2P) network, which is a slightly different kind of distributed system [15] than that illustrated in the example. Now distributed systems are groups of independent nodes connected with others in a specified manner in order to produce a common result, and they are strictly structured in such a way that these groups appear to be a single well-defined system for the end user.

Through these networks, each system can communicate with the others by messages and responses. The main advantage is that communication between each system provides synchronization as well as an error-free environment. Most of the distributed systems are effectively bounded with synchronous messaging channels.

By analyzing each node, the following can be found:

- The nodes are largely programmable, autonomous, asynchronous and failure-free.
- Each node has its own storage and computing processor. They have shared memory and can operate concurrently.

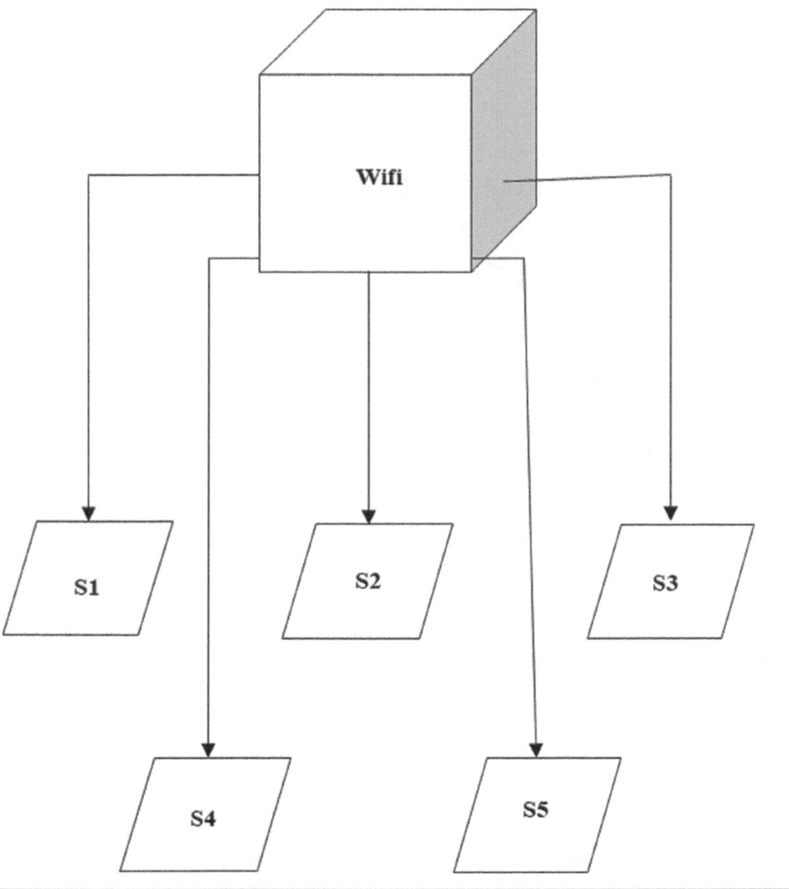

Figure 1.9 Ethernet connections.

- The nodes are interconnected with others to offer services, and share or store data (e.g. blockchain).
- All nodes communicate with others by using messages.
- Every node in the distributed system is capable of sending and receiving messages to and from the others.

1.5.2 Distributed Computing Architecture

There are mainly two types of architecture:

1. Client–server architecture
2. Peer–peer architecture

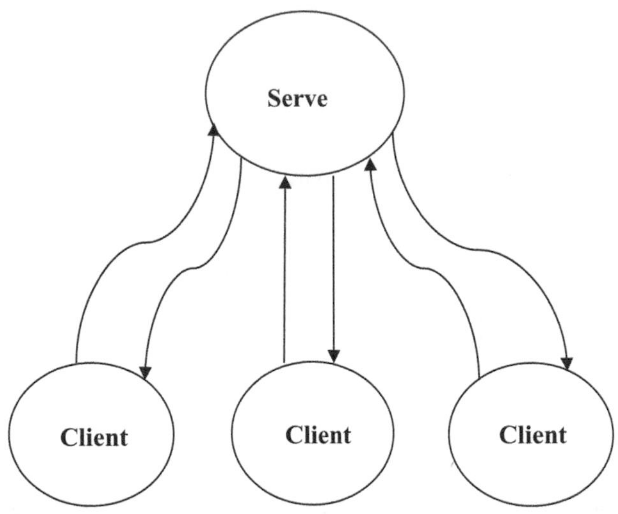

Figure 1.10 Client–server architecture.

1.5.2.1 Client–Server Architecture

In client–server architecture, the main entities are:

1. Server
2. Clients

   Server: An entity that is purely responsible for offering services to the client; servers provide services like storage, data processing, deploying applications, etc.

   Client: A client is an entity that communicates with the server in order to complete its local task. They are normally connected to the server on the Internet (Figure 1.10).

This architecture is a good example of a service-oriented system.

The biggest disadvantage of this type of architecture is that the complete system is dependent on the central single point (server). If the server goes down, then the whole system stops.

In this architecture, there is different layered architecture [16] in which, according to the purpose, several layers can be added on the client side as well the server side in order to accomplish the system requirements, security and complexity. Commonly used layered architecture types include two-tier and three-tier architecture. Each architecture type has its own characteristics to provide maximum security to participants.

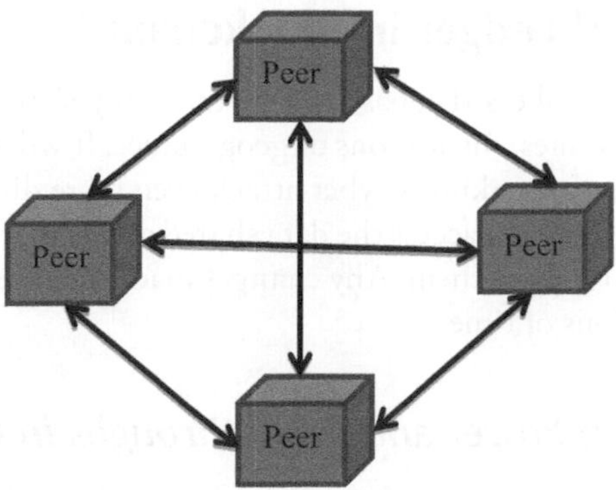

Figure 1.11 Peer-to-peer architecture.

1.5.2.2 Peer-to-Peer Architecture

P2P architecture is a network of inter-connected systems in which they are capable of sharing resources and information. Every system connected to the network is referred to as a node or "peer". This type of architecture can be used in blockchain technology, transportation services, education, e-commerce, banking and finance, etc.

Advantages of P2P architecture are:

1. It can be easily configured.
2. It is easy to install.
3. All the nodes are capable of sharing resources with other nodes and can communicate with other nodes present in the network.
4. If any single node goes down it will not affect the complete system.
5. Maintaining such architecture is comparatively cost-effective (Figure 1.11).

Blockchain technology works on the principle of peer-to-peer architecture; it helps the technology to be more powerful, secure and efficient. Blockchain can be used for many industrial purposes but it is most commonly used in "cryptocurrencies".

A peer-to-peer network is centric when it comes to managing transactions within a blockchain. All the nodes can communicate with others and transact with the others in the blockchain. All peer-to-peer networks are decentralized, and a blockchain is also a decentralized application. This characteristic makes the blockchain technology more secure than other technologies and very hard to hack or break into. But the most complicated part is that backups and security must be provided to each node individually, and there is no centralized entity to manage all the nodes in the architecture.

1.6 Distributed Ledger in Blockchain

A distributed ledger is like a database that is manually shared and synchronized across multiple nodes, sites, institutions or geographies. It will provide transactions with public witnesses, by making a cyber-attack even more difficult. The objects at each node of the network can access the data shared across that distributed network and have an identical copy of them. Any changes made to ledgers are reflected to all other nodes in fractions of time.

1.6.1 Computing Power and Breakthroughs in Cryptography

A distributed ledger of any transactions or contracts is established in decentralized form across different locations and people, eliminating the need for a single central authority to keep a barrier against manipulations. All the data on it are stored securely using cryptographic techniques. Once the data/information are stored, it becomes an immutable database, which is one of the basic rules of the network.

The abstraction at the center of blockchain systems is the notion of a *ledger*, an invention of the Italian Renaissance developed to support the double-way-entry book-keeping system, a distant precursor of modern cryptocurrencies [16]. A ledger is just an indelible, append-only-log of transactions between various parties (Figures 1.12 and 1.13).

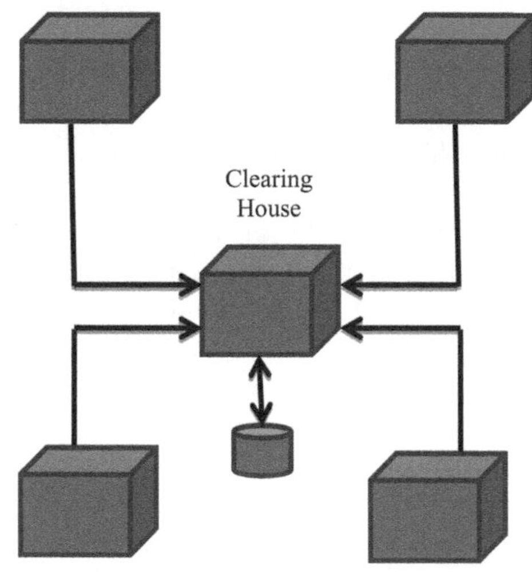

Figure 1.12 Centralized ledger.

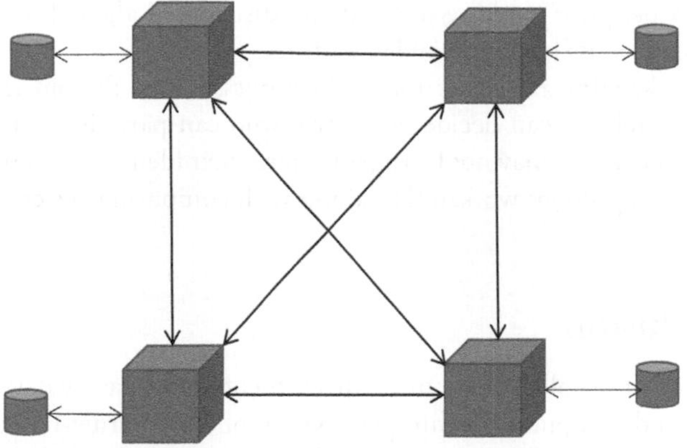

Figure 1.13 Distributed ledger.

Real-world examples of distributed ledgers:

- Government procedures
- Institutions
- Corporate work
- Issuance of passports
- Licenses
- Voting procedures
- Finance
- Agreements
- Identity cards

While the distributed-ledger technology has more advantages, it's at an early stage and is still being explored. The future of centuries-old ledgers is the decentralized ledger.

1.6.2 Public and Private Blockchain

The difference between *private* (*permissioned*) blockchain systems, where objects have reliable and authorized identities and only strictly vetted parties can participate, and *public* (*permissionless*) blockchain systems, where objects cannot be reliably identified and anyone can participate, is critical for making sense of the blockchain landscape.

Private blockchains [17] are better for business applications, particularly in regulated industries, like finance, subject to know-your-customer and anti-money-laundering regulations. Moreover, private blockchains also tend to be better at governance. Most prior working methods on distributed algorithms have focused on system participants having reliable identities.

Public blockchains are better for applications such as Bitcoin [18–19], which guarantee that nobody can decide or control who can participate in the scenario, and participants may or may not be eager to have their identities known. Each node can act as an independent workstation along with computing systems.

1.7 Conclusion

The application of the blockchain mechanism is enhanced day by day with different mechanisms and computing techniques. Many of the internet-based methodologies are now using the advantages of distributed systems as well as the blockchain mechanism. Once submitted any data in the system will be safe forever, which enables many applications in the field of finance and other related sectors. Some of the areas are the following:

VeChain is a blockchain platform which aims to enhance business operations by improving the tracking of products and processes. BitGold is a 2005 proposal that resembles Bitcoin's consensus system and incorporates hashes. A cryptocurrency can be defined as a digital or virtual currency that uses cryptography. A cryptocurrency is very difficult to counterfeit because of this security feature. Digital copy is a duplicate record of every Bitcoin transaction that has taken place over a peer-to-peer network.

References

1. Nagasubramanian, Gayathri, Rakesh Kumar Sakthivel, Rizwan Patan, Amir H. Gandomi, Muthuramalingam Sankayya, and Balamurugan Balusamy. "Securing e-health records using keyless signature infrastructure blockchain technology in the cloud." *Neural Computing and Applications* (2018): 1–9.
2. Westerlund, Magnus, and Nane Kratzke. "Towards distributed clouds: a review about the evolution of centralized cloud computing, distributed ledger technologies, and a foresight on unifying opportunities and security implications." In *2018 International Conference on High Performance Computing & Simulation (HPCS)*, pp. 655–663. IEEE, 2018.
3. Archer, Charles J., Michael A. Blocksome, James E. Carey, and Philip J. Sanders. "Administering virtual machines in a distributed computing environment." U.S. Patent 10,255,098, issued April 9, 2019.
4. Meng, Gang. "Stable data-processing in a distributed computing environment." U.S. Patent 10,044,505, issued August 7, 2018.
5. Wong, Wai Ming, and Michael C. Hui. "Method and system for modeling and analyzing computing resource requirements of software applications in a shared and distributed computing environment." U.S. Patent Application 10/216,545, filed February 26, 2019.

6. Cairns, Douglas Allan. "Efficient computations and network communications in a distributed computing environment." U.S. Patent 10,248,476, issued April 2, 2019.
7. Archer, Charles J., Michael A. Blocksome, James E. Carey, and Philip J. Sanders. "Administering virtual machines in a distributed computing environment." U.S. Patent 10,255,098, issued April 9, 2019.
8. Dillenberger, Donna Eng, and Gong Su. "Parallel execution of blockchain transactions." U.S. Patent 10,255,108, issued April 9, 2019.
9. Li, Keqin. "Scheduling parallel tasks with energy and time constraints on multiple manycore processors in a cloud computing environment." *Future Generation Computer Systems* 82 (2018): 591–605.
10. Chen, Zhen, Pei Zhao, Fuyi Li, André Leier, Tatiana T. Marquez-Lago, Yanan Wang, Geoffrey I. Webb et al. "iFeature: a python package and web server for features extraction and selection from protein and peptide sequences." *Bioinformatics* 34, no. 14 (2018): 2499–2502.
11. Wei, Leyi, Shasha Luan, Luis Augusto Eijy Nagai, Ran Su, and Quan Zou. "Exploring sequence-based features for the improved prediction of DNA N4-methylcytosine sites in multiple species." *Bioinformatics* 35, no. 8 (2018): 1326–1333.
12. Salloum, Said A., Mostafa Al-Emran, Azza Abdel Monem, and Khaled Shaalan. "Using text mining techniques for extracting information from research articles." In *Intelligent Natural Language Processing: Trends and Applications*, pp. 373–397. Springer, Cham, 2018.
13. Shae, Zonyin, and Jeffrey Tsai. "Transform blockchain into distributed parallel computing architecture for precision medicine." In *2018 IEEE 38th International Conference on Distributed Computing Systems (ICDCS)*, pp. 1290–1299. IEEE, 2018.
14. Xiong, Zehui, Yang Zhang, Dusit Niyato, Ping Wang, and Zhu Han. "When mobile blockchain meets edge computing." *IEEE Communications Magazine* 56, no. 8 (2018): 33–39.
15. Puthal, Deepak, Nisha Malik, Saraju P. Mohanty, Elias Kougianos, and Chi Yang. "The blockchain as a decentralized security framework [future directions]." *IEEE Consumer Electronics Magazine* 7, no. 2 (2018): 18–21.
16. Liu, Hong, Yan Zhang, and Tao Yang. "Blockchain-enabled security in electric vehicles cloud and edge computing." *IEEE Network* 32, no. 3 (2018): 78–83.
17. Hughes, Alex, Andrew Park, Jan Kietzmann, and Chris Archer-Brown. "Beyond Bitcoin: what blockchain and distributed ledger technologies mean for firms." *Business Horizons* 62, no. 3 (2019): 273–281.
18. Dr. Kavita. "A future's dominant technology blockchain: digital transformation." In *IEEE International Conference on Computing, Power and Communication Technologies 2018 (GUCON 2018) organized by Galgotias University*, Greater Noida, 28–29 September, 2018.
19. Casado-Vara, Roberto, and Juan Corchado. "Distributed e-health wide-world accounting ledger via blockchain." *Journal of Intelligent & Fuzzy Systems* 36, no. 3 (2019): 2381–2386.
20. Pop, Claudia, Tudor Cioara, Marcel Antal, Ionut Anghel, Ioan Salomie, and Massimo Bertoncini. "Blockchain based decentralized management of demand response programs in smart energy grids." *Sensors* 18, no. 1 (2018): 162.
21. Saugata Dutta, and Dr Kavita. "Evolution of blockchain technology in business applications." *Journal of Emerging Technologies and Innovative Research (JETIR)* 6, no. 9: 240–244, JETIR May 2019.

Chapter 2

Blockchain Components and Concept

M. R. Manu, Namya Musthafa,
B. Balamurugan, and Rahul Chauhan

Contents

2.1 Evolution of Blockchain

The blockchain has evolved since 1991, starting with Stuart Haber and W Scott Stornetta's work on cryptographically secure chain of blocks, the first work on a cryptographically secured block chain where no one tampered with time stamp of document. Then in 1992, the system was upgraded with the Merkle tree approach, which optimized and combined all tasks into a single one. In the year 2008 blockchain gained relevance due to a group of people named Satoshi Nakamoto. Satoshi Nakamoto is the accredited brain behind the digital ledger technology. The new concepts and approaches evolved into the blockchain mechanism for transformation towards digital data utilization in the year 2009. In the beginning it was developed to support Bitcoin. Decentralized data using a decentralized database are the core components of blockchain. The need for Bitcoin increased drastically so blockchain made immediate changes to the Internet. The Russian-Canadian transferred money in form of Bitcoin scripting language.

The decentralized nature of the blockchain mechanism can make any language readable by computer rather than third party, which will generate smart contracts. The Ethereum projects are useful in efficient transaction management systems. The security of transaction through the blockchain methodology produces different digital transaction systems such as Bitcoin, cryptocurrency, Ethereum, and light coin ripple which can handle huge numbers of transaction per second.

In the years 2013–2015 the system developed to Ethereum development with version blockchain 2.0. It provides for the recording of books as well as contracts. This can develop the decentralized application efficiently. In the year 2018 a new version of blockchain evolved: blockchain 3.0. It supports the leveraging capabilities of blockchain. The new blockchain application is called NEO, which is an open source platform first developed in China. For further upgrades with the Internet of Things, IOTA was developed. It supports the Internet of Things ecosystem for digital transactions.

2.1.1 Architecture of Blockchain

The blockchain provides a peer-to-peer distributed ledger mechanism for transaction management in a secure manner. Each ledger is a block which is interlinked to other blocks in the structure. The databases are shared among each other in a distributed way. There is a timestamp server for controlling the databases and each block is associated with a reference to the previous block. This reference is also managed by a hashing mechanism for security (Figures 2.1 and 2.2).

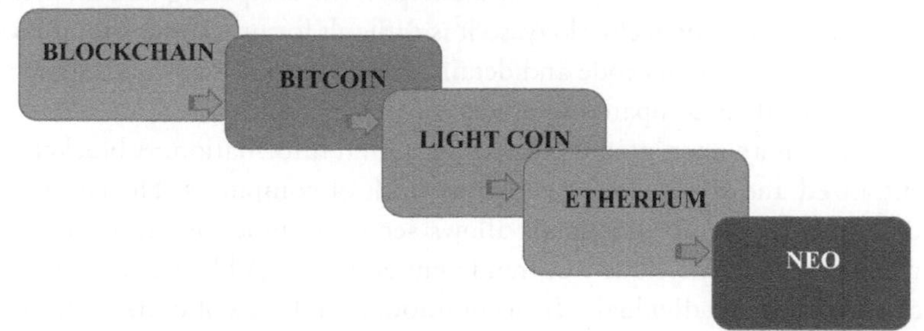

Figure 2.1 Evolution of blockchain.

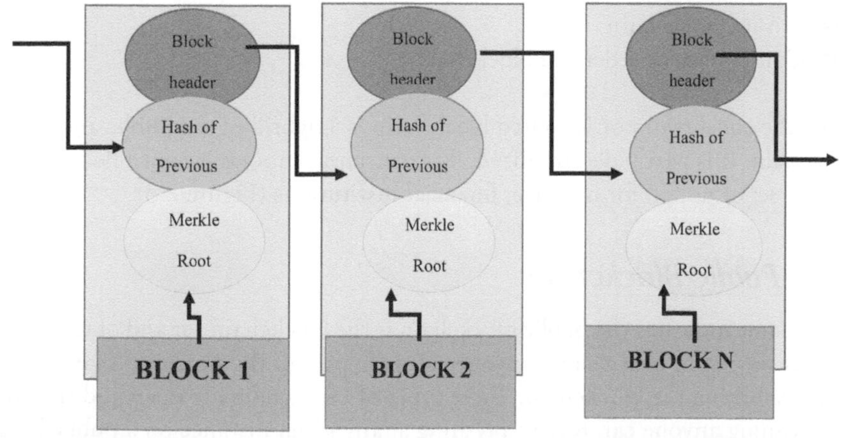

Figure 2.2 Structure of blockchain.

2.1.2 Blocks in Blockchain

Blocks are the basic units of the blockchain. A block is a basic data structure for transaction distributed to other monetary control. The blocks contain a block header which verifies the validity of the block. It contains metadata which describe the block. The metadata information of a block are mentioned below:

a) **Version filed**: Which describes the current version of the block.
b) **Previous block header hash:** References the previous block's parent block.
c) **Merkle root**: Cryptographic hash of all transactions involved in this block.
d) **Nonce and nbits**: The number of times the process repeated so that it becomes a complex task

2.2 Types of Blockchain

The concept of cryptocurrencies (Bitcoin) introduced the idea of blockchain into the spotlight; it is a database that protects the data from tampering and analysis. The blockchain is still a rising technology, so it is difficult for us to understand its working without getting into its code and details. Blockchain is a new and more secure connected network as compared to others.

Blockchain is an encrypted repository of digital information. A blockchain has a decentralized and distributed style of network of computers. Hence its hosting on a distributed network of systems allows secure transactions to occur across a blockchain with little possibility of fraudulent activities. A blockchain allows users to track assets across individuals. To accommodate all kinds of users there are three major types of blockchains. The three types of blockchains are:

- Public blockchain
- Private blockchain
- Consortium or federated blockchain

Note: The consortium or federated blockchain is a hybrid of the public and private blockchain. It is partly decentralized. The consensus process is controlled by a preselected set of nodes, for instance, financial institutions (Figure 2.3).

2.2.1 Public Blockchain

As the name indicates the public blockchain is the blockchain for and of the public. There is no one in charge, and anyone can take part in the processes like reading/writing/auditing the blockchain. These types of blockchain are open and transparent, meaning anyone can review anything at any given instance on a public blockchain. This idea will raise the question that if no one is in charge of anything here

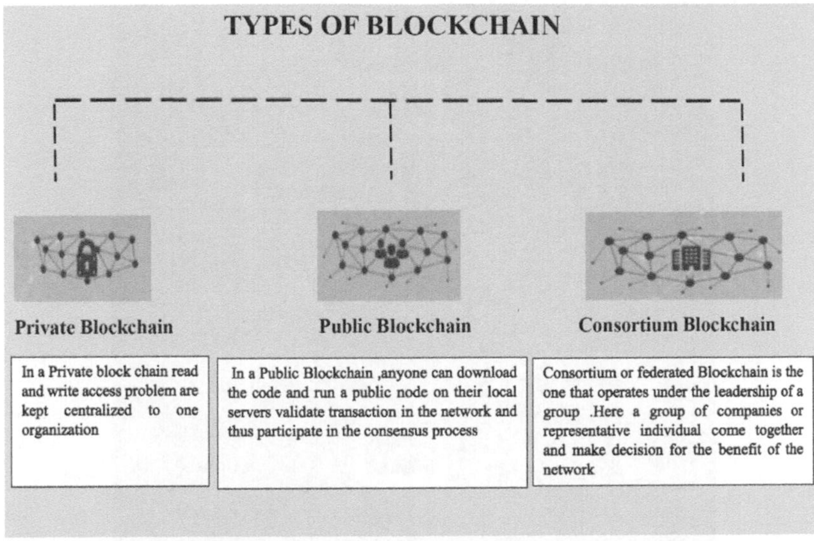

Figure 2.3 Types of blockchain.

then how are the decisions taken on these types of the blockchain? It is done by any of a variety of decentralized consensus mechanisms. Here are some of examples for decentralized consensus mechanisms:

- Proof of work (PoW)
- Proof of stake (PoS)

There are three things we have to be aware of, that make a public blockchain really public. They are as follows:

- The code to operate a public blockchain is openly available so that anyone can download the code and start running a public node on their local device, validating transactions in the network and participating in the consensus process. This gives anyone the right to participate in the process that determines which blocks get added to the chain and what the current shape and size of the blockchain is.
- Anyone can be part of transactions in the network. Hence the transactions should go through as long as they are valid.
- Anyone can access and read transactions using a block explorer. Transactions are transparent but anonymous (Figure 2.4).

From the figure we can understand that anyone can participate in a public blockchain, without permission.

Examples of public blockchains include Bitcoin, Ethereum, Monero, Dash, and Litecoin, among others.

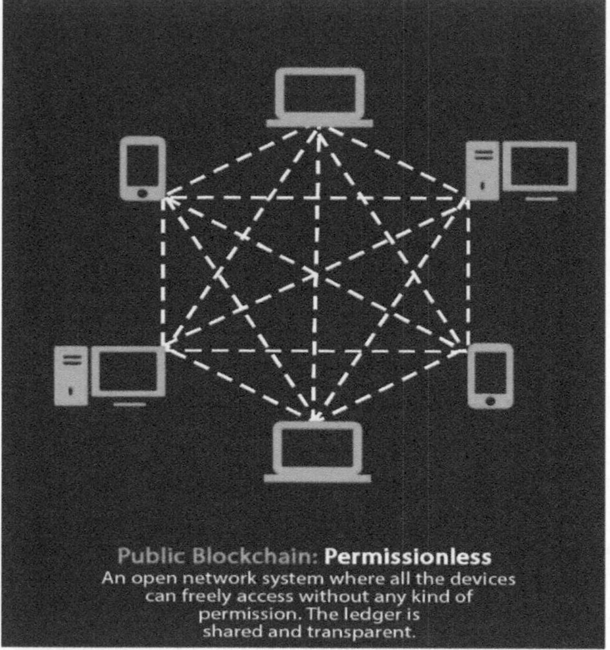

Figure 2.4 Public blockchain.

On Bitcoin and Litecoin blockchain networks

- Anyone can run BTC/LTC full node and start mining.
- Anyone can make transactions on the BTC/LTC chain.
- Anyone can review/audit the blockchain in a blockchain explorer (Table 2.1).

The nature of the public blockchain leads to two major implications.

1. Everyone can potentially shatter current business models through the reduction in the use of intermediaries.
2. By using a blockchain, we do not necessarily have to maintain servers or have system administrators. Hence we can minimize the cost of creating and running decentralized applications or DApps.

2.2.2 Private Blockchain

A private blockchain as its name indicates is a private asset of an individual or an organization. Unlike a public blockchain, a private blockchain has an in-charge who monitors important tasks such as read/write or whom to selectively give access to read or vice versa. A private blockchain is also known as a permissioned blockchain as it has restrictions on who can access it and also who can participate in transaction and validation. Only previously chosen entities have permissions to access the

Table 2.1 Public vs. Private vs. Consortium Blockchain

Public Blockchain	*Private Blockchain*	*Consortium or Federated Blockchain*
Anyone can run BTC/LTC full node	Not everyone can run a full node	Selected members of the consortium can run a full node
Anyone can make transactions	Not everyone can make transactions	Selected members of the consortium can make transactions
Anyone can review/audit the blockchain	Not everyone can review/audit the blockchain	Selected members of the consortium can review/audit the blockchain
Examples: Bitcoin, Litecoin, etc.	Example: Bankchain	Examples: r3, EWF

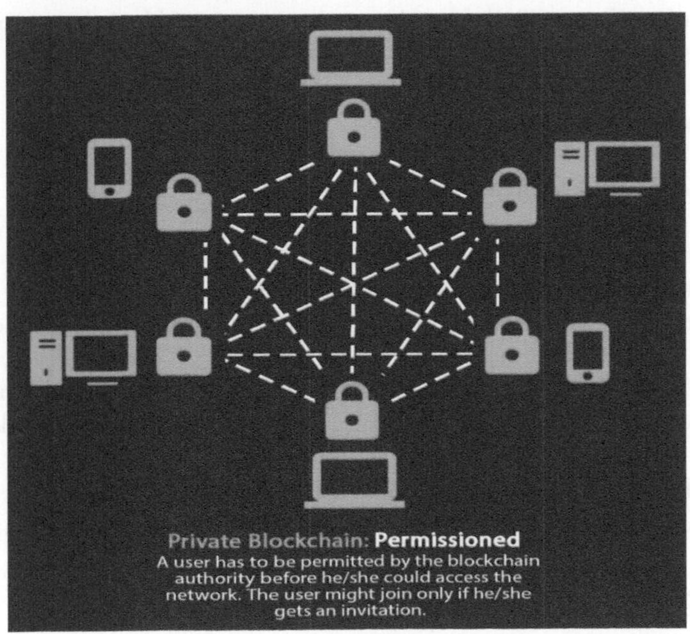

Figure 2.5 Private blockchain.

blockchain. How are these entities are chosen? It is done by the respective authority and are given permission by the chain developers while building the blockchain application. An important consensus is achieved on the whim of the central in-charge who can give mining rights to anyone or not give them at all. Suppose there is a need to grant permissions to new users or revoke permissions from an existing user, then it can be taken care of by the network administrator (Figure 2.5).

Private blockchains are mostly used in database management and auditing, among other fields. Their uses are internal to a single company, and so the companies will not want the data to be accessible to the public. They use blockchain technology by setting up groups and participants who can verify transactions internally.

However, private blockchains may scale better and comply better with the security and privacy regulations of governmental data. A private blockchain runs the risk of security breaches just like in a centralized system. Thus, they have certain security advantages, and other security disadvantages, like a coin has two sides. Blockchain is still in the emerging stages, so it is conjecture how this ground-breaking technology will evolve and be adopted. Some examples of private blockchains include MONAX and Multichain.

The important advantages of private blockchains are minimal transaction costs and data redundancies as well as easier data-handling and more automated compliance functionalities. That's what makes it centralized again where various rights are exercised and vested in a central trusted party but yet it is cryptographically secured from the company's point of view and more cost-effective for them. But it is still debatable if such a private thing can be called a 'blockchain' because it fundamentally defeats the whole purpose of blockchain that Bitcoin introduced to us.

- Example: Bankchain

In such types of blockchain:

- Not everyone can run a full node and start mining.
- Not everyone can make transactions on the chain.
- Not everyone can review/audit the blockchain in a blockchain explorer.

Similarly as we observed for public blockchains, here also we can encounter some key implications of the implicit nature and characteristics of private blockchains.

- Reduction in transaction costs and data redundancies
- Simplified data-handling and more automated compliance mechanisms

2.2.3 Consortium or Federated Blockchain

A consortium blockchain is like a hybrid of public and private blockchains. In this type of blockchain, some nodes control the consensus process, and some other nodes may be allowed to participate in the transactions. In other words, this type of blockchain can be used when organizations are ready to share the blockchain, but restrict data access to them, and keep it secure from public access. That is, it possesses the characteristics of a public blockchain as the blockchain is being shared by different nodes, and also it behaves like a private blockchain by restricting the access to the blockchain from the different nodes. Therefore, it is partly public and partly private.

A consortium blockchain consists of two types of users. They are:

1. The users who have control over the blockchain and decide who should have permission to access the blockchain
2. The users who can access the blockchain

Here instead of a single authority in charge, you have more than one in charge. Basically, you have a group of companies or representative individuals coming together and making decisions for the benefit of the whole network. Such groups are also called consortiums or a federation, hence the name consortium or federated blockchain.

For example, let's suppose you have a consortium of the world's top 20 financial institutes; you have decided in the code that only if a transaction or a block or decision is voted/verified by more than 15 institutes then it should get added to the blockchain. So it is a way of achieving things much faster, and you also have more than one single point of failures which in a way protects the whole ecosystem against a single point of failure (Figure 2.6).

- Example: r3, EWF

In such a blockchain:

- Members of the consortium can run a full node and start mining.
- Members of the consortium can make transactions/decisions on the chain.
- Members of the consortium can review/audit the blockchain in a blockchain explorer.

2.3 The Logical Components of Blockchain

Cryptocurrencies are the technology that is built on blockchain to enable a shared distributed tamper-proof ledger to be viewed by anyone with the corresponding software. Unleashing blockchain technology from its application to cryptocurrencies is very important in understanding the broader implications and applications of blockchain technology. Differentiating the two will be helpful in understanding why there is such excitement about blockchain-inspired ruptures. Bioinformatics, governance, banking, trading, society, politics, and even the very structure of the Internet itself are suited for disruption. Generally, blockchain technology will bring disintermediation among everything.

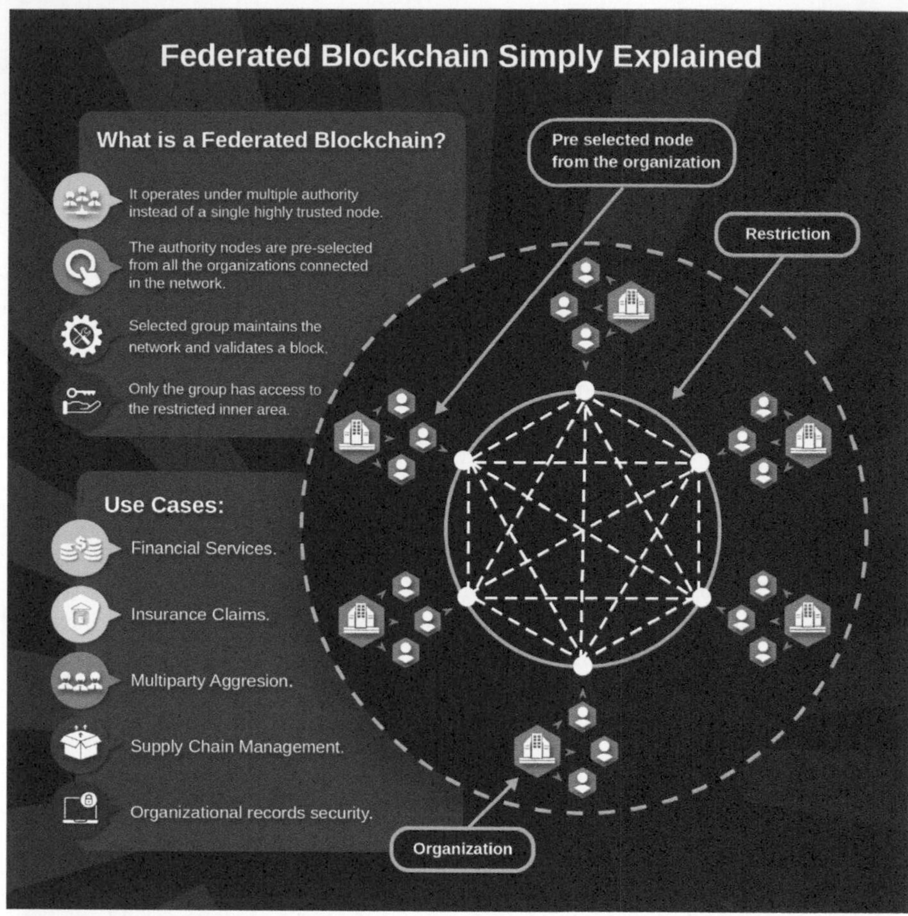

Figure 2.6 Federated blockchain.

To understand blockchain technology applications deeply it is necessary to understand the logical components of a blockchain ecosystem and the duties of each component. The four main components of any blockchain ecosystem are given below:

- A node application
- A shared ledger
- A consensus algorithm
- A virtual machine

1. **Node Application**

   Each computer inter-connected through the Internet needs to install and run a computer application specific to the ecosystem they desire to participate in. For example using the case of Bitcoin as an ecosystem, each computer must be running the Bitcoin wallet application. In some blockchain applications, like Bankchain, participation is restricted and requires special permissions to join (referred to as permissioned blockchains). Bankchain only permits banks to run the node application. But in the Bitcoin ecosystem anyone can download and install the node application and also participate in the ecosystem.

2. **Shared Ledger**
   The distributed ledger is a data structure managed inside the node application. Once you have the node application running, you can view the respective ledger (or blockchain) contents for that ecosystem. Interaction is done according to the rules of the ecosystem in which it resides. You can run as many node applications as you like and are permitted to use, and each will participate in their respective blockchain ecosystems. It is important to note that the number of ecosystems you are a participant in doesn't matter as you will only have one shared ledger for each ecosystem.
3. **Consensus Algorithm**
   The consensus algorithm is implemented as a portion of the node application, by providing the 'rules of the game' for how the ecosystem will arrive at a single view of the ledger. Different ecosystems have different methods for attaining consensus depending on the desired features of the ecosystem. Participation in the consensus-building process, the method for determining the 'world state' of the ecosystem, can be vested in a number of different schemes: proof-of-work, proof-of-stake, proof-of-elapsed-time; each method qualifies nodes as honest in a different way before participation in the consensus-building process.
4. **Virtual Machine**
   A virtual machine is a representation of a machine (real or imaginary) created by a computer program and operated with instructions embodied in a language. It is an abstraction of a machine, held inside a machine. To some degree we are already are accustomed to the abstraction of real-world objects and entities as virtual objects in a computer. Think of a button in a graphical user interface of an application. You press the button on the screen and the state of the program inside the computer changes. Another example might be your driver's license as it is represented in a government computer. It is an abstraction of your real-world legal authorization to operate a motor vehicle, and it is largely what counts these days, rather than the real-world physical printed driver's license.

2.4 Core Components of Blockchain Architecture

- **Node** – user or computer within the blockchain architecture (each has an independent copy of the whole blockchain ledger)

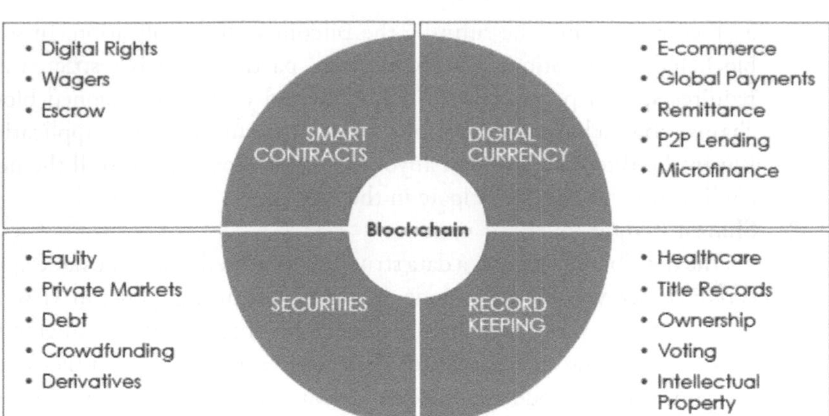

Figure 2.7 Blockchain applications.

- **Transaction** – smallest building block of a blockchain system (records, information, etc.) that serves as the purpose of the blockchain
- **Block** – a data structure used for keeping a set of transactions which is distributed to all nodes in the network
- **Chain** – a sequence of blocks in a specific order
- **Miners** – specific nodes which perform the block verification process before adding anything to the blockchain structure
- **Consensus** (consensus protocol) – a set of rules and arrangements to carry out blockchain operations (Figure 2.7)

2.4.1 Ledger Management

The blockchain is the underlying technology behind technology like Bitcoin. A distributed ledger is essential as it is a list of all events and transactions entered onto it and is held simultaneously by each node in the network. Whenever a new event or transaction is added to the ledger, encryption is done to everything; by adding to the ledger, the task becomes complex. The ledger is both visible to everyone in the network and also secured so that people can't tamper with it. Every new piece of information added to this ledger is added as a 'block'. This block is mathematically encrypted and is approved to be added to the ledger according to a series of consensus protocols, that is, ways of approving additions and protecting against fraud or double spending without the need for a centralized authority (Figure 2.8).

A distributed ledger is a database that is decentralized as it is distributed across several distinct computers or nodes. Here every node will maintain the ledger, and if any data changes happen, the ledger will get updated. The updating takes place independently at each node. Through the ledger along with a little of computer code, you can create 'smart contracts'. These are a series of clauses which are added

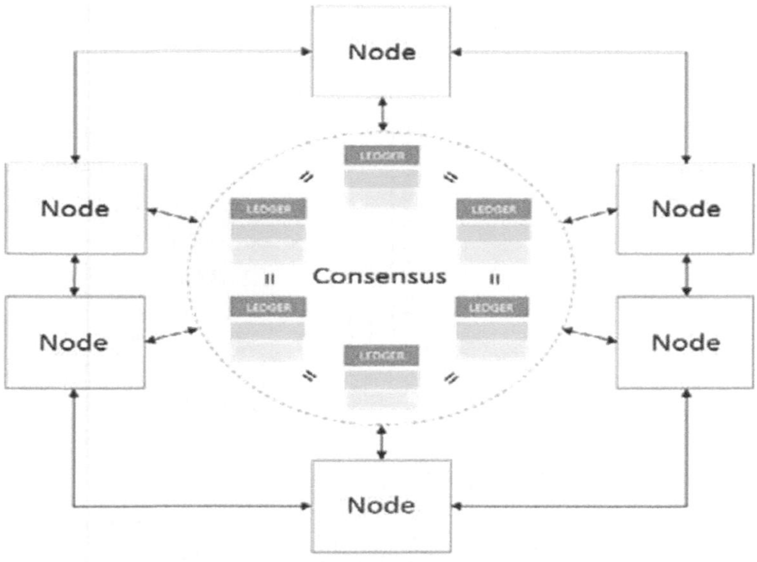

Figure 2.8 Ledger management with consensus.

to the ledger and powered by computer code. When the clause in the ledger is met, the computer code activates and the next step of the contract is triggered.

All the nodes are equal in terms of authority. There is no central authority or server managing the database that makes the technology transparent. Every node can update the ledger, and other nodes will verify its existence. This property of distributed ledgers makes them an attractive technology for the financial industry or any other industry looking for more transparent technology and those who need technology which is far from central authority.

By using distributed ledgers, there is no need for centralized authority. It is a network of ledgers or contracts that is maintained by nodes. The nodes that can be merged into blocks which make it even easier to maintain larger distributed network ledgers. Even without a central authority, all the information stays secure. To enable the distributed network, technology such as cryptography is required to assign the data with cryptographic signatures and keys for use. Anything that is stored on the distributed ledger is immutable. Immutability makes it even harder for hackers to try to hack distributed ledger networks such as Bitcoin. Additionally the absence of a central authority means that it is also free from any intentional change as well (Figure 2.9).

Three major steps are involved:

- To initiate a payment, entity A digitally signs a proposed update to the shared ledger with cryptographic tools, to transfer funds from its account on the ledger to entity B's account.

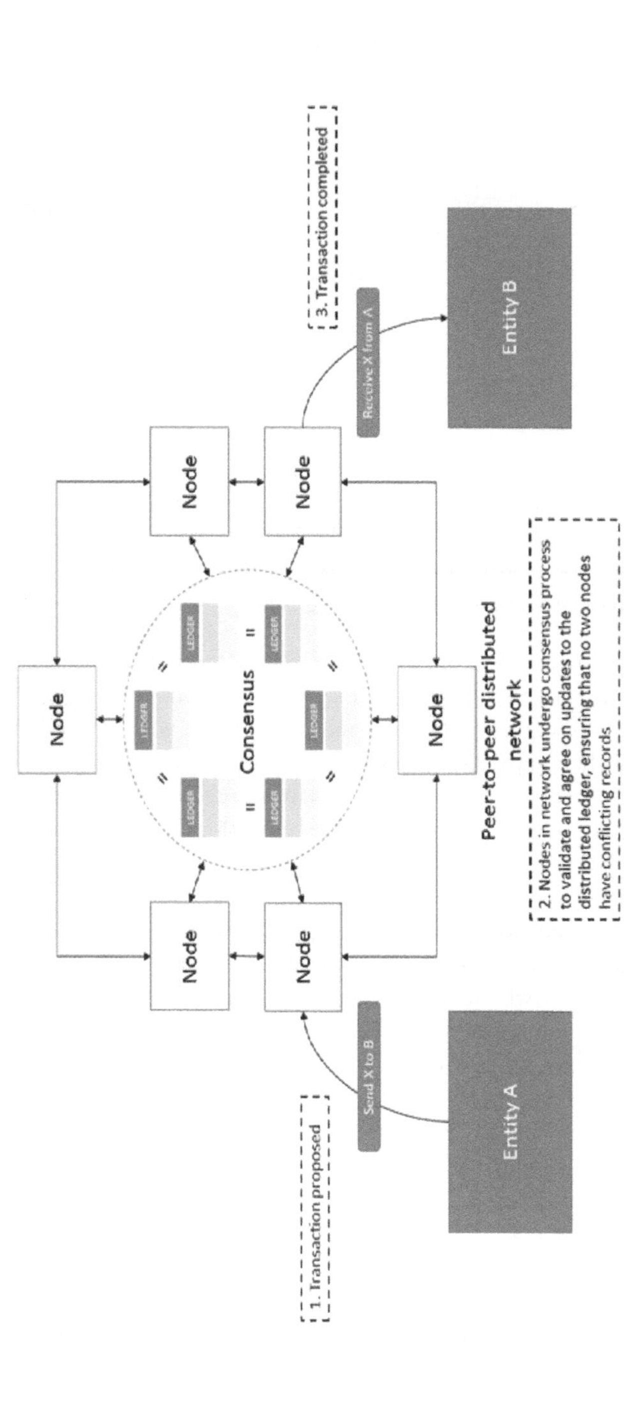

Figure 2.9 Ledger management in distributed network.

- Upon receiving the transfer request, other nodes authenticate entity A's identity and validate the transaction by checking to make sure that entity A has the necessary cryptographic credentials to make an update to the record in question. Validation would include, among other things, verifying that entity A has sufficient funds to make the payment. Nodes also take part in the consensus process to agree on the payments that should be included in the next update to the state of the ledger.
- After the update has been accepted by the nodes, the properties of the asset are modified such that all future transactions regarding the asset must be initiated using the cryptographic credentials of entity B.

2.5 Smart Contracts and Their Applications

Smart contracts are translations of an agreement consisting of terms and conditions into a computational code, which is script. The developers write the script in a programming language like Java, C++, etc. in such a way that it lacks ambiguity and does not lead to interpretation. It is a mechanism that involves digital assets and two or more parties. And here some or all of the parties deposit assets into the smart contract, and the assets automatically get redistributed among those parties according to a formula based on certain data, and it is not known at the time of initializing a contract.

A smart contract is a set of lines of code that is uploaded and stored to check a contract's validity and containing a set of rules under which the parties who share the smart contract agree to interact with each other. It is automatically executed when previously determined and defined terms and conditions are met. The smart contract code facilitates, verifies, and enforces the conference or performance of an agreement or transaction. It is the simplest form of decentralized automation.

The smart contract is defined and executed inside a distributed blockchain. And also every transaction and contract execution should happen inside the blockchain. There are certain steps to enable smart contract execution:

- During the coding procedure, blockchain developers write the smart contract as a script using a programming language, and also implement the logic behind the contract so that when a given action or transaction occurs, the script enables the following step.
- Once the contract is coded completely, the script is sent out to the blockchain. Execution of the code is done by a distributed network. Typically those computers that are already made available for computation can execute the contract, and the output of this execution should be the same for identical input regardless of the computer where it is executed.
- Several conditions can be coded, and final smart contract users may choose the conditions that are needed for their smart contract.

Execution of contract is run in a peer-to-peer manner; it is relatively similar to decentralization. Simple users connected to the Internet may act as clients; hence, they have to install the client in the computer. We refer to this principle as mining, and the computers used to run the program are called nodes.

The main difference between a smart contract and a traditional contract is that the former doesn't rely on a third party; cryptographic code enforces it. At the initial level, these are programs that run according to how the developers have set them up to run. For an example we can consider a vending machine where we are going to implement a smart contract mechanically. It verifies the following characteristics:

- No third-party involvement in the transaction
- When a coin is inserted into the machine and a product is selected, it delivers the product directly to us as long as the terms and conditions are met; here the condition is: Our coin has the same or higher value than the product that we want to purchase.

Benefits of smart contracts

The benefits of smart contracts go hand-in-hand with blockchain.

- Speed and accuracy: Smart contracts are digital and automated, so you won't have to spend time processing paperwork or reconciling and correcting the errors that are often written into documents that have been filled manually. Computer code is also more exact than the legalese that traditional contracts are written in.
- Trust: Smart contracts automatically execute transactions following predetermined rules, and the encrypted records of those transactions are shared across participants. Thus, nobody has to question whether information has been altered for personal benefit.
- Security: Blockchain transaction records are encrypted, and that makes them very hard to hack. Because each individual record is connected to previous and subsequent records on a distributed ledger, the whole chain would need to be altered to change a single record.
- Savings: Smart contracts remove the need for intermediaries because participants can trust the visible data and the technology to properly execute the transaction. There is no need for an extra person to validate and verify the terms of an agreement because it is built into the code.

Advantages:

- The cost is minimized by removing intermediaries.
- Contract execution time is reduced; every action is executed automatically according to coded rules.

- Automatic process: A third party is not involved to enable contract.
- By removing intermediaries, the cost of money transfers can be lowered.
- It uses a transparent system: Anyone can have access to the blockchain.
- Protects data and transaction from fraud. It is impossible to change or update the data inside a blockchain and still maintain a coherent chain.
- The decentralization aspect prevents the system from collapse which is the case when a centralized system is down.

2.6 Applications of Smart Contracts

Smart contracts also allow for more complex transactions to be carried out between two anonymous parties without the need for a central authority, enforcement system, or legal guidance. This means a smart contract can be programmed to enable a wide variety of actions. Smart contracts can be used to allow an entire world of new applications designed to solve many real-world problems.

Consider an example of a poet who allows journalists and content creators to control and manage their digital rights. This means they maintain the ability to offer up their content on an open marketplace for a fee, or enter into agreements with clients and be paid as work is completed. No escrow service, lawyers, or agencies are required.

As you can imagine, there are many industries that can benefit from this kind of technology, such as:

- Intellectual property
- The legal industry (contracts, negotiations, etc.)
- Shipping and logistics
- Finance/banking
- Real estate

2.6.1 Financial Services and Insurance

One of the important challenges facing the insurance industry regularly is fraudulent activities. For insurance companies to overcome this problem there needs to be an administrative team that looks into claims and ensures their validity. Smart contracts regulate the impact of this major challenge because both the insurer and insured can show a bond with each other by an agreement without the use of notaries, lawyers, and other intermediaries. This cost-saving opportunity would ultimately be passed down to the end consumer. While this doesn't inherently prevent fraud, it can help prevent arguments in court. However, a blockchain acting as a public ledger and system of record combined with the benefits of a smart contract would make it much more difficult to slip under the radar.

2.6.2 Mortgage Transactions

Another important application of smart contracts is in the mortgage industry. Blockchain technology can allow for buyers and sellers to be automatically connected together in a friction-less, hassle-free process. Construct a smart contract governing all terms and conditions – evade the need for lawyers, realtors, and other professionals. This saves both time and money for both sides of the transaction while also minimizing any potential errors or costs that could otherwise come from doing things manually.

2.6.3 Supply Chain Transparency

Tracking packages as they move around the world is a difficult task, but smart contracts can simplify it. From the moment a product leaves the factory floor to when it arrives on store shelves, the transparent nature of this technology can make the entire process more simple; it clearly shows where exactly every package is along with where in the supply chain potential errors take place. For example, in the case of a contaminated shipment, management will be able to see exactly where each individual product came from and isolate the contaminated goods without throwing away an entire shipment. Not only does this help organizations save costs, but also it keeps buyers safer.

2.6.4 Medical Research

As researchers in the medical field conduct clinical trials and research potential cures to diseases such as cancer, effectively sharing data amongst the various institutions freely and openly is something smart contracts can facilitate. Data can be freely exchanged without compromising the privacy and data security of the patients and subjects involved. A smart contract consists of various if–then scenarios that work well in this particular use case.

2.6.5 Digital Identity and Records Management

Although in the current era, huge technical companies get away with mining our data and personal information, in the future this could change severely with the use of smart contracts. Individuals can own and control their own digital identity, including passwords, data, digital assets, records, and other details. This would be really different from our current situation, where often dozens of different institutions, organizations, and parties all have their own individual copies of our personal information – an obvious security risk. Instead, all these details can be consolidated and owned by an individual who chooses with whom to share this information with smart contracts.

2.6.5.1 Peer Network and Membership Management

A blockchain network is in the style of a peer-to-peer network which is running a decentralized blockchain framework. As we know a network includes one or more members, who have unique identities in the network. For example, a member might be an individual or an organization in a consortium of banks. Each member runs a single or multiple blockchain peer nodes to run chain-code, endorse transactions, and store a local copy of the ledger.

Consider the case of Amazon Managed Blockchain, that creates and manages these components for each and every member in a network, and also creates components that are shared by all members in a network, such as the Hyperledger Fabric ordering service and the general networking configuration. The user can choose different editions of Amazon Managed Blockchain according their wishes and requirements. This edition determines the capacity and capabilities of the network.

The creator also necessary should create the first Managed Blockchain network member. Additional members are added through a proposal and voting process. There is no need to pay for the network establishment, but each member pays an hourly rate (billed per second) for their network membership. Charges in each network vary depending on the edition of the network. Each member also pays for peer nodes, peer node storage, and the amount of data that the member writes to the network.

The blockchain network remains active as long as there are members participating in it, and the network is deleted only when the last member deletes itself from the network. No member or AWS account, even the creator's AWS account, can delete the network until they are the last member and delete themselves.

2.6.5.2 Inviting and Removing Members in a Peer Network

Initially an AWS account creates a Managed Blockchain network, but surprisingly this network is not owned by that AWS account, or any other AWS account. Hence a Managed Blockchain network is decentralized. To alter the configurations of the network, members make proposals that all other members in the network vote on. If another AWS account desires to join the network, an existing member creates a proposal to invite the account. Other members can vote Yes or No on the proposal. If the proposal is approved by gaining Yes as majority an invitation is sent to the AWS account. The account then accepts the invitation and creates a member to join the network. Similarly when a member in a different AWS account is required to be removed a proposal for removal is submitted. A principal in an AWS account with sufficient permissions can remove a member that the account owns at any time by deleting that member directly, without submitting a proposal for voting.

The voting policy for the network is defined by the network creator when they create the network. This voting policy determines the basic rules, such as the percentage of votes required to pass the proposal, and the duration before the vote expires, etc., for all proposals voting on the network.

Whenever a new member joins the network, one of the first things they must do is create at least one *peer node* in the membership. Blockchain networks contain a distributed, cryptographically secure ledger that maintains a history of transactions in the network that is immutable—it can't be changed. Each peer node stores a local copy of the ledger in a distributed manner. Each peer node also holds the global state of the network for the channels in which they participate, that gets updated with each new transaction performed in the network. The peer nodes also interact with each other to create and endorse the transactions that are proposed on the network. Based on their business logic and the blockchain framework being used members can define the rules in the endorsement. In this way, every member can independently verify the transaction history without a centralized authority.

2.7 Applications and Implementation of Blockchain

By now, we all know the working of blockchain. Blockchain creates a ledger of transactions which is secure, tamper-proof, and can be easily accessible. Like the Internet, blockchain has no central authority, instead it is a network of transactions shared over a vast network of users. It is made up of a chain of blocks, where each block holds data, the hash code of the block, and the hash of the previous block. If data on any block are changed, then its hash code changes due to which the next block no longer points to the previous block.

There are six basic steps of blockchain where each step represents different aspects of blockchain:

1. First, a transaction of some sort is defined. The transaction can be a literal transaction, such as a user wanting to send money to another user, or can be less literal, such as a user trying to pass a secure token for identification.
2. Second, this transaction is codified into a block, which is then added to the network for processing.
3. Third, the block is presented to all distributed members, and is compared amongst them for integrity and, in some cases, against previous records of ledgers to prove authoritatively whether or not it is valid.
4. Fourth, the members within the blockchain either deny or approve the block itself.
5. Fifth, the block is either denied or approved and if approved, the block is added to the chain of records.
6. In the sixth and final step, the transaction is approved and carried out. In the case of a financial transaction, the money changes hands; in this case, it is like a token; the token generated is then verified by network and trusted throughout the network.

Due to these features, blockchain has a huge number of applications. Some of the important applications of blockchain are discussed.

2.7.1 Blockchain Technology in Finance

Blockchain for finance is not a very new concept. But the idea of blockchain became famous when Satoshi Nakamoto used the concept of blockchain in his idea which he called Bitcoin. Bitcoin is a perfect example of blockchain for finance.

There is no arguing that blockchain got its recognition after Bitcoin. Many other currencies came into the market when Bitcoin start getting famous. We called these currencies cryptocurrencies because they use cryptographic functions such as hash 256, etc.

Here's a good example of what a single cryptocurrency block may look like:

```
block = {
    'index': 1,
    'timestamp': 1506057125.900785,
    'transactions': [
        {
            'sender': "8527147fe1f5426f9dd545de4b27ee00",
            'recipient': "a77f5cdfa2934df3954a5c7c7da5df1f",
            'amount': 5,
        }
    ],
    'proof': 324984774000,
    'previous_hash': "2cf24dba5fb0a30e26e83b2ac5b9e29e1b161e5c1fa7425e73043362938b9824"
}
```

The blockchain is a very efficient and fail-proof method of verifying a transaction. It is virtually impossible to fake a transaction in cryptocurrencies like Bitcoin.

To fake a transaction in Bitcoin, one has to alter all the blocks on a blockchain on every copy of the blockchain. Blockchain also uses a concept called proof of work. Due to the concept of proof of work, a user has to solve a problem which requires a high amount of processing power, and only that person who solves the problem first is allowed to add a new block in the blockchain. The amount of cost and labor it saves for the global financial market is so appealing that many of the famous and major financial institutions have already started to invest millions of resources to research how best to implement it.

2.7.2 What Can Blockchain Do for the Financial and Banking Industries?

Blockchain has the potential to fully change the financial services that we are using today. Some of the top factors that blockchain can change in finance are:

2.7.2.1 Fraud Detection

Blockchain is getting bigger because it can handle fraud detection in a way that our normal banking system can't. Most of the banking system nowadays is a centralized system. There is one server that holds records of all the transactions. But these systems are vulnerable to cyberattacks. If any hacker breaches the system, then he has full access to make any fraud. The blockchain is essentially a distributed ledger. In the blockchain, each block contains a timestamp and holds some batches of individual transactions. These records also contain a link to a previous block. It is believed that this technology has the power to eliminate some of the current crimes that are being committed online today against our financial institutions.

2.7.2.2 Know Your Customer (KYC)

According to Thomson Reuters Survey, financial companies spend anywhere between \$60 million to \$500 million to keep up with know your customer (KYC) and customer due diligence regulations. Blockchain would allow the autonomous confirmation of one client by allowing one organization to be reached by other companies so the KYC process wouldn't have to start over again.

2.7.2.3 Payments

Blockchain division could be highly transformative in the payments process. Blockchain will give higher security and lower costs to organizations like banks to process payments between companies and their buyers and even between banks themselves. In the current reality, there are a lot of intermediaries in the payment processing system, but blockchain would reduce the need for a lot of them.

2.7.3 Problems in Implementing Blockchain in Financial Services

Blockchain provides huge numbers of opportunities to improve our financial services but before implementing this there are some hurdles which we need to clear first.

The blockchains that would be used by financial institutions would need to comply with privacy laws of today and the future and need to ensure the safety of the data. There are many questions regarding regulatory oversight for this new technology that need to be sorted out. And any blockchain used in this sector would need to handle an extraordinarily large data set, therefore scalability is incredibly important.

2.8 Blockchain Security in Online Voting

Another application for blockchain technology is in e-voting systems. We all know the importance of fair elections in a democratic country where altering votes is a serious crime. There are many tools for fair elections such as electronic voting machines (EVM), but cyberhackers have already proved that they are vulnerable. In a world of rapidly increasing technological progress, old technology such as EVM can no longer provide 100% insurance against changes in votes. Blockchain is fool-proof technology and what is once written is not easy to change forcefully. This feature of blockchain makes it perfect for voting. By casting votes as transactions, we can create a blockchain which keeps track of the tallies of the votes. This way, everyone can agree on the final count because they can count the votes themselves, and because of the blockchain audit trail, they can verify that no votes were changed or removed, and no illegitimate votes were added. A similar blockchain-based mobile voting system had been scheduled to be used in the U.S. state of West Virginia's midterm elections, Cointelegraph reported September 27. Following the early November elections, the state's Secretary of State noted that 144 military personnel stationed overseas from 24 counties were able to successfully cast their ballots on a mobile, blockchain-based platform called Voatz.

In May, Cointelegraph released an analysis of the potential usage of blockchain technology for elections.

This year, several countries worldwide have announced the consideration of blockchain-based systems for voting, such as Ukraine, Catalonia, and the Japanese city of Tsukuba. Back in June, the Swiss city of Zug, commonly known as 'Crypto Valley', conducted a blockchain-powered trial municipal vote, as Cointelegraph wrote June 9.

Blockchain systems not only have higher security than traditional voting systems but also open the doors to online voting or e-voting. We all know that blockchain cannot be altered; there is already some research going on in which people are trying to perfect e-voting systems.

2.8.1 Challenges in E-Voting Applications for Blockchain

The challenge in this sector is not security or the money required to change the current system; the main reason we are still using old system is that a resilient and inclusive voting system should be something a citizen or group of citizens can understand and replace by pen and paper if it fails. Not everyone is on the Internet; it is true that many still struggle to vote though EVM, and these technologies are still new for a huge number of people in developing countries like India. The system must remain backward-compatible and have realistic fallback options, for something so sensitive

2.8.1.1 Blockchain-Based Certification

Blockchain can also be used in creating authentic certificates or to verify the authenticity of certificates. Counterfeiting in certificates has been a longstanding issue. Not until the Massachusetts Institute of Technology Media Lab released their project of Blockcerts, a technique which is mainly implemented by conflating the hash value of local files to the blockchain, did an effective technological approach protecting authentic credential certification and reputation appear, but there remain numerous issues.

Based on Blockcerts, a series of cryptographic solutions have been proposed to resolve the issues above, including utilizing a multi-signature scheme to ameliorate the authentication of certificates exerting a safe revocation mechanism to improve the reliability of certificates revocation establishing a secure federated identification to confirm the identity of the issuing institution.

2.8.1.2 Working Insight

When a user presents a document, the technology converts or encodes the document into a cryptographic digest or cryptographic hash. Satoshi Nakamoto's white paper on Bitcoin carries a permanent hash of

> b1674191a88ec5cdd733e4240a81803105dc412d6c6708d53ab94fc
> 248f4f553.

Submitting the same document more than once, for verification, will have the hash and the transaction markers match each time. If the document contains any changes, the markers won't match. The user will also have the power to allow or disallow said organization or individual from viewing the document.

2.8.1.3 How to Verify Documents on a Blockchain

Currently, there are multiple ways in which one can verify the existence of a document on the blockchain. The easiest of them is to re-upload the document to verify its existence. Upon re-uploading of the document, the proof of its existence is verified, as the cryptographic digest and the marker for the transaction are also verified. The other ways are to check the transaction record of the Bitcoin blockchain to verify the existence of a time-stamped document. Returning to the verification page of the original time-stamped document also verifies its existence. Thus, the existence of a time-stamped document on a prior date is proven.

This will help banks, educational institutions, and healthcare industries verify documents with much less time, cost, and effort.

2.9 Building a Blockchain

Before starting, remember that a blockchain is an immutable, sequential chain of records called blocks. They can contain transactions, files, or any data you like, really. But the important thing is that they're chained together using hashes.

Building a blockchain is not something for which you require a degree; anyone with basic programming knowledge can create their own blockchain.

2.9.1 Generations of Hashes

By now you may know that each block of a blockchain contains a hash value, which is changed if someone tries to change a single work in of the page. Each block also contains a hash of the previous block and the next block, So, if someone tries to change any value, he/she has to change the hash on every block on the blockchain. The hash is what makes the blockchain foolproof.

Before starting to build blockchain, we need to get some basic knowledge of what hashing is in cryptography.

Hashing is generating a value or values from a string of text using a mathematical function.

In simple words, hashing means taking a string of variable size and converting it into an output of fixed length. Cryptocurrencies like Bitcoin use Secure Hashing Algorithm 256, also known as SHA-256.

Let's see how the hashing process works. We are going to put in certain inputs. For this exercise, we are going to use the SHA-256.

Data

Hello World!

SHA-256 hash

7f83b1657ff1fc53b92dc18148a1d65dfc2d4b1fa3d677284addd200126d9069

As you can see in the image, hashing can take input of any length and generate fixed output (256-bit output in SHA-256).

It is important to note that a particular string always generates the same hash output. This hashing method is also used in storing passwords; instead of storing the passwords of a user in a database, companies store the hash of the password and whenever the user inputs their password, they compare the hash generated from the user's entered text and the hash stored in the database; if both hashes match, the user is logged in.

Another important thing about hashes is that hashing is a one-way function. That means if anyone has the hash key of a string, it is impossible to generate that string back from its hash value.

2.9.2 Let Us See How You Make Your Own Hash Function in Python

Let us see an example by creating your own hash function in Python. I am assuming you already know how to install and use Python on your system.

First, open Python on your system. You can do this by going to terminal and typing Python.

This command will put you into the Python REPL, an environment where you can try out Python commands directly as opposed to writing a program in a separate file.

Then, type the following. Don't ignore the tab.

```
import  hashlib
def
hash(mystring):
       hash_object = hashlib.md5(mystring.encode())
       print(hash_object.hexdigest())
```

You have now created a function, hash(), which will calculate and print out the hash value for a given string using the MD5 hashing algorithm. To run it, put a string in between the parentheses in quotation marks, e.g.:

```
hash("AnyString")
```

After that, press ENTER to see the hash digest of that string.

You will see that calling the hash function on the same string will always generate the same hash, but adding or changing one character will generate a completely different hash value:

```
hash("AnyString") =>
7ae26e64679abd1e66cfe1e9b93a9e85 hash("AnyString!")
=> 6b1f6fde5ae60b2fe1bfe50677434c88
```

In the Bitcoin protocol, the hash functions are a major part of the block hashing algorithm which is used to write new transactions into the blockchain through the mining process.

2.10 API Creation for Blockchain

One of the best elements of the blockchain is the fact it is driven entirely by the concept of trust. Each interaction on the blockchain trusts and verifies the transaction, and depends on the consensus of all nodes to track what is an otherwise untrackable, decentralized activity. The API community is likewise driven by trust as a key concept – and this is why the blockchain is an amazing element of connectivity in the API stack.

2.10.1 So, What Is API?

API stands for application programming interface. It is a way for software to exchange functionalities between each other. API is a software intermediary that allows two applications to talk to each other (Figure 2.10).

2.10.2 How to Integrate Blockchain APIs in a Website

There are hundreds of different APIs available on the Internet which give huge functionalities to us. In this guide, we shall consider one of the simplest applications for setting up a blockchain API. An API allows Bitcoin payments to be accepted on a website. The process depends on Receive Payments API V2 of blockchain to generate new unused addresses to receive payments for a specific extended public key (xPub):

- First, to make a request for an API key, users must set up a wallet at www.blockchain.info and request an API key at https://api.blockchain.info/v2/apikey/request/.
- The next step is to generate an extended public key which is also called xPub. If you generate a wallet from the address given above, the xPub can be found in:

  Settings → Addresses → Manage → More Options → Show xPub

- Now we generate a unique address for each customer. The basic URL for creating a new request for every customer is: https://api.blockchain.info/v2/receive?xpub=$xpub&callback=$callback_url &key=$key.

This is an API key which contains three parameters. These parameters are as follows

1. xpub – your xPub.
2. callback_url – a callback URL which is to be notified when a payment is received.
3. key – your blockchain.info API key which you have created in step 1. Please note that every call to the server will increment the index by one to avoid showing the same address to different customers.

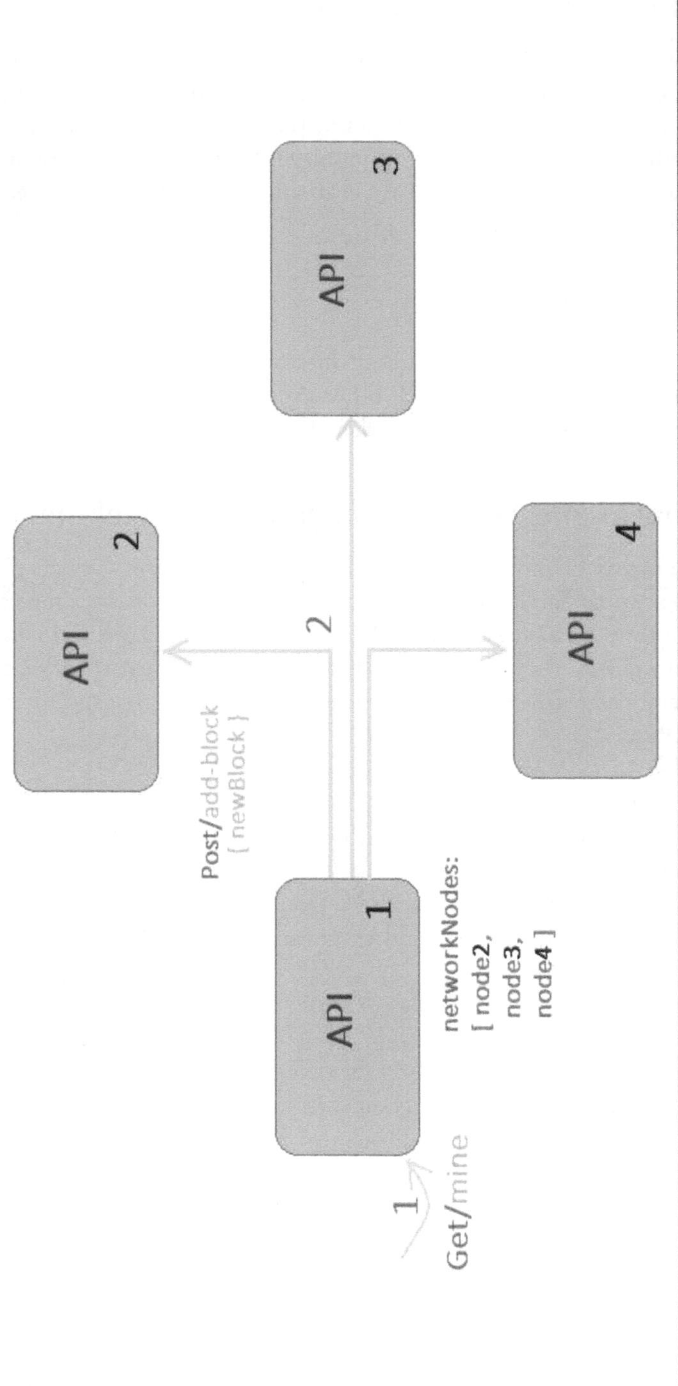

Figure 2.10 API creation of blockchain.

References

W. Akins, J. L. Chapman, and J. M. Gordon, "A whole new world: Income tax considerations of the bitcoin economy," 2013. [Online]. Available: https://ssrn.com/abstract=2394738.

A. Biryukov, D. Khovratovich, and I. Pustogarov, "Deanonymisation of clients in bitcoin p2p network," In *Proceedings of the 2014 ACM SIGSAC Conference on Computer and Communications Security*, New York, NY, USA, 2014, pp. 15–29.

Blockchain Wikipedia. Available: https://en.wikipedia.org/wiki/Blockchain 2: Bitcoin Wikipedia. Available: https://en.wikipedia.org/wiki/Bitcoin.

I. Eyal and E. G. Sirer, "Majority is not enough: Bitcoin mining is vulnerable," In *Proceedings of International Conference on Financial Cryptography and Data Security*, Berlin, Heidelberg, 2014, pp. 436–454.

Follow my vote. Available: https://followmyvote.com/online-voting-technology/blockchain-technology/ 5: cointelegraph.

G. Foroglou and A.-L. Tsilidou, "Further applications of the blockchain," In *12th Student Conference on Managerial Science and Technology*. 2015.

G. Hileman. "State of blockchain q1 2016: Blockchain funding overtakes bitcoin," *Coindesk*, 2016. [Online]. Available: http://www.coindesk.com/state-of-blockchain-q1-2016/.

A. Kosba, A. Miller, E. Shi, Z. Wen, and C. Papamanthou, "Hawk: The blockchain model of cryptography and privacy-preserving smart contracts," In *Proceedings of IEEE Symposium on Security and Privacy (SP)*, San Jose, CA, USA, 2016, pp. 839–858.

S. Nakamoto, "Bitcoin: A peer-to-peer electronic cash system," 2008. [Online]. Available: https://bitcoin.org/bitcoin.pdf.

C. Noyes, "Bitav: Fast anti-malware by distributed blockchain consensus and feedforward scanning," arXiv preprint arXiv:1601.01405, 2016.

B. Marr. "Practical examples of how blockchains are used in banking and the financial services sector," *Forbes*, 2017. Available: https://www.forbes.com/sites/bernardmarr/2017/08/10/practical-examples-of-how-blockchains-are-used-in-banking-and-the-financial-services-sector/#23adfc4c1a11.

G. W. Peters, E. Panayi, and A. Chapelle, "Trends in crypto-currencies and blockchain technologies: A monetary theory and regulation perspective," 2015. [Online]. Available: http://dx.doi.org/10.2139/ssrn. 2646618 563.

T. K. Sharma, "Documentation verification using blockchain" *Blockchain Council*, 2017. Available: https://www.blockchain-council.org/blockchain/document-verification-system-using-blockchain/.

M. Sharples and J. Domingue, "The blockchain and kudos: A distributed system for educational record, reputation and reward," In *Proceedings of 11th European Conference on Technology Enhanced Learning (EC-TEL 2015)*, Lyon, France, 2015, pp. 490–496.

M. Yakubowski. "South Korean government to test blockchain use for e-voting system," *CoinTelegraph*, 2018. Available: https://cointelegraph.com/news/south-korean-government-to-test-blockchain-use-for-e-voting-system.

Y. Zhang and J. Wen, "An iot electric business model based on the protocol of bitcoin," In *Proceedings of 18th International Conference on Intelligence in Next Generation Networks (ICIN)*, Paris, France, 2015, pp. 184–191.

Chapter 3

Blockchain and IoT Security

D. Peter Augustine and Pethuru Raj

Contents

3.1 Overview

The fashion of technology merged with the resourceful invention evolving in the information world in the current scenario is blockchain, which is the breakthrough of a person identified by the alias Satoshi Nakamoto in 2008. But since then, blockchain has gone through radical growth with the assured support in every dimension of applications in the information technology world. Meanwhile the question in the minds of everyone who comes across the word of blockchain is "what is blockchain?"

Blockchain has changed the view of the Internet by making the information distributed but not copied. Even though the main intention in the invention of blockchain was for the digital currency, Bitcoin, currently there are many more effective and efficient uses of the technology that have been explored by the IT world.

Blockchain can be defined as a *decentralized, distributed and public digital transaction ledger with the idea of recording those transactions across numerous systems; in turn, any record of transactions cannot be edited with effect from a date of creation, without the modification of all succeeding blocks.*

A common man can define blockchain in the simplest form as *records of data flowing in with series of time-stamps which cannot be changed and which are organized by a set of computers, not possessed by any sole unit.* The time-stamped record in the series can be viewed as a block, and by means of cryptographic algorithms, the blocks are fully bound to one another. Hence, the blocks and the chain of the interlinked blocks lead to the term blockchain.

Emerging blockchain technology has had a vibrant influence, disrupting the way IT has been working, because the blockchain network does not possess any central control. Subsequently it enables a shared and immutable ledger; the information carried in the record can be accessed by anyone. Henceforth, the knowledge of any domain put up on the blockchain will be in the public domain, and anyone involved in any time-stamp in the series is responsible for their activities.

3.2 Understanding Blockchain

A blockchain requires only infrastructure costs as in any other case. But it does not require any transaction cost. The underlying mechanism in blockchain is simple, secured, effective and robust transactions between any two systems by automated means. It can be viewed as a transaction between any two parties. In this scenario, the first party namely "A" initiates and requests a transaction from the other party "B". The initialization of the transaction is followed by the creation of a block. This block is broadcasted to all the computers distributed in the network. The block is verified by all the systems in the network and validated. This verified and validated block is added to a chain and the chain is stored across the network; the birth of new exclusive record with a unique history. The transactions are done with verification and executed between "A" and "B". Any scenario which affects a single record falsely will in turn incorrectly affect the entire chain coupled with the blocks in millions of instances.

An algorithm for a transaction between any two parties using blockchain:

User A requests a transaction with B.
Block is created to represent the transaction.
Block is broadcasted to all the systems in the network.
Validation of the block is done by all the nodes.
Validated block is added to the chain.
Verification of the transaction is done and executed between A and B.

Let us have an example of a railway system where millions of transactions are executed every day. Customer can buy a ticket online either through the app or through the website. Usually the credit or debit card service providers demand a processing fee for their services. Using the blockchain technology, railway systems can eliminate the processing fee and even enhance the efficiency of the entire process. In this picture the railway system can be seen as "A" and the customer can be viewed as "B". The ticket can be an individual block of a record which is validated and added to the chain. A ticket is autonomously verifiable and immutable like any financial transaction one carries out in day-to-day life. The entire transaction, including the particular train route, train network, each ticket sold and each journey traveled in the route, can be viewed as a record. The people involved in this transaction to update or modify by any means are completely responsible for the transaction.

Even smart homes, using the Internet of Things, which produces data from different sensors, video cameras and any other connected smart home appliances, can be coupled with blockchain technology for the utmost security of data and processes. This will be pondered over in the last section of this chapter.

One can understand that blockchain can give better cost savings by avoiding the middle man. In the case of Ola, let us imagine the adaptation of blockchain technology which is a big threat for the middle man who is charging additional fees. This is because the middle man can removed just by encoding the information of the transaction occurring. This can lead to the complete avoidance of agents coming in the middle, who play a vital role in manipulating the costs in different sectors. This is thanks to the technology of blockchain which cuts the cost for everyone involved in the transactions in a network.

There is no transaction cost in blockchain technology. So the cost incurred or the profit earned is absolutely between the end parties alone and there is no middle-man involvement in any way.

3.3 Towards an Optimized Blockchain for IoT

Before beginning with the optimized blockchain for IoT, one can have an overview of the IoT and why it needs blockchain technology for its effectiveness.

In the simplest form, we can define IoT as the things which are cable of sending and receiving data in the world of the Internet for controlling the device or analyzing and manipulating the data shared. IoT begins with home appliances such as lights, fans, refrigerators, televisions, etc., and broadens to any electronic device in the network. IoT works with a combination of sensors as the input medium, software to control it and the network for passing data between any objects in the network. It meets purposes such as data analysis, cost cutting or predictive analysis depends on the objective of the implementation.

In this IoT boom, the big giants in the manufacturing sectors such as Samsung, Siemens and the IT giants like IBM and AT&T are working on the best adaptation and usage of IoT for automation from the basic predictive maintenance stage to the high-end data analytics level.

Indeed IoT gives a greater range of advantages in the fields of smart cities, smart grids and healthcare. Amidst all these countless benefits, there is are serious privacy concerns raised by the incessant acquisition of volumes of data from the devices for processing and analyzing. Let us remember the object-oriented programming concept of data encapsulation. So in this IoT world, we can understand that the spontaneous flow of incredible volumes of data from numerous devices may lead to the leakage of data or unauthorized access of data if it does not provide guaranteed security mechanisms. We can visualize some of the privacy challenges like lack of central control, heterogeneity in device resources, multiple attack surfaces, context-specific risks and scale.

As discussed earlier in this chapter, blockchain technology can be used as an effective tool for the protection of privacy and security of data in the IoT era. Blockchain security primarily originates from a cryptographic puzzle identified as proof of work (PoW), used for adding new blocks into the chain. The unfixed public key for the users' identities is used by blockchain to render a high level of privacy. The salient features of blockchain and its usage in various non-financial applications can be used for providing distributed privacy and security in IoT.

3.3.1 Proof of Work

This is the most popular algorithm being used by currencies such as Bitcoin and Ethereum, each one with its own differences.

Even though we can utilize most of these features, blockchain needs to be optimized according the IoT. So the adoption of blockchain by IoT is not straightforward. The following are some major challenges that must be addressed:

(i) There is an actor to be elected as a leader and the block to be chosen to get added to the blockchain in proof of work. In such a situation there is a need to solve a particular mathematical problem to find a solution where resource requirements are high.

(ii) Miners are the actors with the computing power for most of the time due to their first solvable capability. In this scenario, scalability issues originate from the need to achieve consensus among the miners.
(iii) Even if it is not IoT, in the case of cryptocurrency, great deferrals are endorsed to PoW and mechanisms to avert twofold expenditure.

We can examine the scenario of smart home optimization to understand the aforesaid challenges since we have a case study of the same at the end of this chapter. Some researchers have proposed different optimization techniques with respect to the above-mentioned challenges. One of them is a lightweight instantiation of blockchain without compromising the privacy and security benefits extracted from blockchain implementation. They have focused on the optimization of resource ingestion and increasing the network scalability by adopting a hierarchical structure. The structure for optimization comprises a smart home, overlay network and cloud storage as three different tiers.

The transactions held among the IoT devices in the smart home are stored in a private immutable ledger (IL), which can be visualized as a blockchain but with the difference of central management and symmetric encryption to lessen the processing overhead. Another difference in the case of blockchain is that the devices requiring greater resources mutually create a distributed overlay that instantiates a public blockchain. The transactions are the communications held between the entities in the mentioned different tiers which are assembled into blocks. These blocks are attached to the chain without resolving the PoW, which reduces the appending overhead considerably. The verified and validated transactions are accessible for the complete network instantaneously. This mechanism potentially decreases the delay of IoT transactions, like data access or queries. The possible salient distributed trust method can be used in the overlay to decrease the processing overhead in authenticating new blocks.

Even supposing the suggested mechanism has been constructed with the idea of smart home IoT, it can be viewed and tested for similar applications and extended to different IoT applications.

3.4 Blockchain – Backbone of IoT

As we have discussed earlier in this chapter, IoT has taken a place in the technology world incorporating cloud-centric frameworks. The cloud giants like IBM, AWS and Azure follow cloud-based IoT design to portray the analytics done in the cloud platforms through the communication of devices to the cloud. Everyday there is an extensive growth in the number of applications and together IoT devices are also piling up. It will increase 25 and 50 billion in the near future as stated by a report of Gartner and Cisco. But cloud-based IoT suffers from issues of latency, bandwidth and connectivity since the computing takes place at the center of the network. The industrial IoT (IIoT) applications for smart cities, smart factories, smart grids and smart farms are not compatible with cloud-based architectures due to the aforesaid issues. Therefore there is paradigm shift towards edge computing-based application development which may be well-suited for such IIoTs. Edge computing coupled with IoT can avoid all the issues of latency, bandwidth and connectivity. In this scenario blockchain can render a great deal of security support for the IIoT along with edge computing too.

There are concerns about the capability of IoT to protect the limitless number of devices associated to a network. We can understand this from the evidence provided by Accenture which conducted an IoT security overview. It says that "*The urgency for viable IoT security solutions grows by the day. At front-of-mind for many businesses and government leaders lies the same, nagging question: What do we need to do to secure the IoT?*"

There are two major weaknesses to be fixed due to the exponential growth of usage of IoT devices.

1. The first and foremost is security.
2. Meanwhile IoT needs to enlarge its competences to guarantee fast, consistent connections continuously. Because if your Amazon Echo momentarily drops connectivity it may not be a serious issue, but we can imagine how it would be if a self-driving car lost its signal even for an instant.

In the meantime, the blockchain comes onto the scene as a standby to render a great deal of support to overcoming these fundamental issues. The most prized cryptocurrencies in the world are being safeguarded by the blockchain. The entrenched smart contracts and distributed network of the blockchain are substantial solutions for IoT's safety apprehensions. While centralized servers are vulnerable to the security risks which demands more security, the risks are spread out in the network since blockchain works based on distributed computing. But the mechanism of the distributed ledger concept and the continuous connectivity of blockchain guarantees that an issue in one area will not have any influence in any other area. The blockchain can confirm that IoT devices maintain the availability of the connection that they need to provide the services that they offer.

As previously mentioned, let us look at the situation of a driverless car interconnected using the concept of IoT. There is a private blockchain used in this scenario that facilitates safe and concurrent communications from the car initiating with car startup, status verification and smart agreements to exchange the information regarding the insurance and repairs service statistics and real-time locality data to track security.

We can see how the blockchain with distributed ledger technology bridges the serious breaches in IoT in the above context.

1. A blockchain with a distributed ledger offers the confidence, authenticated data of proprietorship, clarity and the complete distributed communication which acts as the backbone for IoT in such distinctive circumstances as mentioned earlier.
2. In the scenario of a smart driverless car system, smart contracts with the insurance companies can be handled effectively and efficiently using blockchain in a secured manner.
3. Each transaction on the blockchain has a timestamp, and each will be protected and maintained for future reference at any time.
4. In the fast-growing current trend of using blockchain in IoT, the developers of IoT will deploy a private blockchain of their own where the transactions can be put in storage for their explicit applications. The current scenario of storing the data on central data servers will become capable of writing into native ledgers that will synchronize with other native ledgers to preserve a sole, but still protected replica of the fact.
5. As a final point with respect to the blockchain as a backbone of IoT, the greatest challenge lies in security of communications in IoT.

3.5 Security Implications of Blockchain IoT

Because of the lack of availability of most of the devices in IoT, the capacity of their availability in network, the variety of devices connected, and the lack of adaptation of safety standards are great challenges. The non-compliance to the standards of the devices will contribute more to the data security breaches. It may lead to unauthorized access to data from one's private space. Some researchers are working on a policy checking method based on the context in the central hub of the network to be incorporated before accommodating and restarting the session of transaction. On the other hand, one cannot eliminate the likelihood of gaining access directly to smart gadgets by bypassing the central hub.

Some other researchers have proposed a safety management provider having dynamic policies based on the content which may restrict access to the devices and the data involved in the transactions. Simultaneously it must include security measures while personal data is allowed to be accessed.

Some researchers have looked into the privacy concerns in smart homes, where the private data very close to an individual is transferred with respect to the sensors' usage in the premises and information gathered.

Some application developers have forwarded the idea of limiting the data access to systems outside the network. We could see some suggestions that the data can be merged with some cryptographic mechanisms to protect the data. At the same time this mechanism may add overhead in different levels, and it may not be necessary in some cases too.

At the outset we can conclude that three major issues need to be resolve with respect to security and privacy in IoT.

1. Resource optimization: The devices constrained with resources may not be appropriate for advanced and composite safety procedures.
2. Privacy: Privacy protection while exposing different types of information.
3. Centralization: It may not be suitable for IoT to use centralized mechanisms since it may add heavy overhead at one end due to high amounts of data.

The novelty in blockchain is fundamentally securing the IoT transitions with a safe information architecture to take the data and authenticate it. The following three major characteristics reveal the implications of blockchain coupled with IoT.

3.5.1 Better Safety of Data

Yes, it's indeed the need for complete protection of data compliance with all the necessary standards without affecting the originality of the data. Since blockchain protocols are not meant to store big data as they are not databases, the provision of "control points" in the blockchain can allow complete control over data access in every level of change.

3.5.2 Robust Structure Creation

The major shortcoming of the current IoT architecture is "having a right and robust structure to share the data across the network" which is overcome by a blockchain which certainly creates the right and robust structure to share the data. Another huge challenge is inter-industry data transactions which may cause more concerns. Blockchain can resolve this greater concern, by simple and easy validation, and authentication.

3.5.3 Implementation of Distributed and Parallel Computing

The challenging implementation of parallel computation along with distributed processing is generally attributed to artificial intelligence applications. On the other hand, it is achieved through blockchain which authenticates and validates each node involved in the networks. Enterprises such as Golem, Hadron, Hypernet, DeepBrain Chain, iExec, Onai, etc., are attempting to solve these problems.

3.6 Blockchain Technology for Large-Scale IoT Systems

The era of data science has allowed anyone in the field of information technology to use data extensively for analysis and making efficient predictions. The other well-advanced area of IT called cloud computing and its advancements also joins hands for improvement in the research field of data science. The big data generated from IoT using cloud-based architecture involving big data analytics will reap the maximum benefits of business solutions. In this context making all possible electronic devices communicate with each other in the name of IoT floods data into the network which may be a local area network, a wide area network or any type of cloud platform. Let us consider this scenario of coupling these technologies of IoT, cloud computing and big data analytics for enterprise solutions. Indeed, it is very true that 100% security is expected where one can never find a pitfall to lose the business at any cost. Figure 3.1 illustrates eight areas across which blockchain enables IoT.

The increasing requirements of IoT and the applications being built on it in association with other cutting-edge technologies make IoT a large-scale system. The cloud is widely used as a central repository for such large-scale IoT systems and the centralized server methodology cannot be a prudent solution. Most IoT systems that are implemented as of now rely on the centralized server concept. But the truth is that these IoT systems use the central server through a network with or without wired Internet. It is necessary for the large-scale IoT systems in view of business development to accomplish the analysis that requires high processing capabilities which is not realized in the existing infrastructure. It is evident that having decentralized or distributed network systems can enhance the existing Internet infrastructure to tackle the huge data processed in

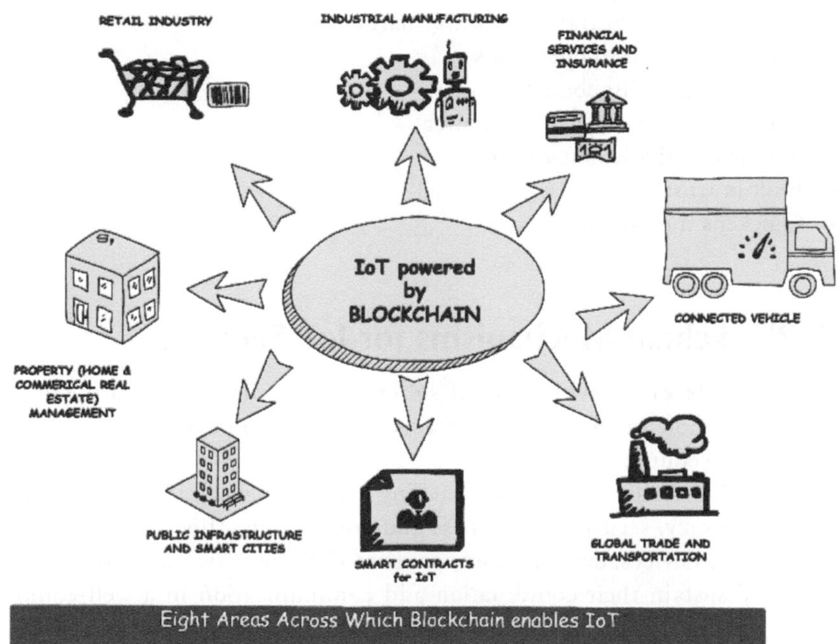

Figure 3.1 Eight areas across which blockchain enables IoT. (From Chemitiganti, V. 2016. What blockchain can do for the Internet of Things. *Vamsi Talks Tech.* Available at http://www.vamsitalkstech.com/?p=3314.)

large-scale IoT systems. The peer-to-peer networking (PPN), distributed file sharing (DFS) and autonomous device coordination (ADC) functions which make the infrastructure efficient can be implemented to track the vast quantity of linked and networked devices in co-ordination.

The blockchain because of its robustness can make IoT systems more reliable and maintain their confidentiality with respect to data. The transactions are also made faster between peers using blockchain due to its distributed ledger.

When IoT is coupled with blockchain the data flow takes the path from sensors, network, router, Internet, distributed systems, blockchain and then for analytics and finally to the end user. There is no possibility of misreading or wrong validations in data in the distributed ledger because of the tamper-resistant nature of blockchain.

The following are some of the most important advantages of blockchain technology:

- Data protection
- Safe communication between peers
- Robust
- Extremely reliable
- High level of data privacy
- Maintaining complete records of actions
- Storing data of old transactions in smart devices
- Allows self-directed operations
- File sharing in distributed manner
- No room for solitary control power
- Removing middleman and cost
- Built-in trust
- Quickens transactions

3.7 Blockchain Mechanisms for IoT Security

There are various ways in which the IoT suffers with the challenges relating to security. Let us see how the blockchain sustains pertaining to intrusion detection and prevention procedures.

Blockchain because of its record-tracking ability can work as a catalytic agent, enhancing privacy, security, robustness and consistency. This can keep track of any number of devices, even millions, connected in the networks of the IoT ecosystem, and sustain their coordination and communication in a well-equipped manner.

The world's first search engine of interconnected devices named "Shodan" supports the user in finding the pitfalls of any IoT device and unveils its vulnerability for restructuring. Using blockchain in any IoT ecosystem will certainly boost the trustworthiness by absolutely eradicating some single point of failure. The hashing algorithms along with cryptographic mechanisms are used in blockchain, to encrypt the data which can yield the best security mechanisms while binding it with IoT. Meanwhile, the flip side of the IoT with respect to the demand of high computing power while using hashing methods and cryptographic techniques is a challenge when combining the blockchain technology with IoT. But the research is going on to overcome these issues at the blockchain level itself.

Some researchers like Underwood see blockchain as the healing remedy for the digital economy where complete security is ensured. Blockchain is considered as the complete tamper-proofing of data in the world of IoT where trust takes first and foremost place.

Nasdaq in October, 2015 itself came out with "Nasdaq Linq" with the blockchain concept to record its transactions with reserved safeties. Because of the advantages of blockchain, Depository Trust & Clearing Corporation (DTCC), USA, is also working with Axoni for the monetary payment facilities such as post-trade stuffs and trades. The government regulatory boards are also keen on blockchain for its capability to render safe, secluded, distinguishable simultaneous watching of trades.

Safeguarding the ongoing operational technology is also of dominant prominence. Therefore, the blockchain can preclude damaging information by handling and safeguarding industrial IoT gadgets. In this wonderful fully protected scenario, once a sensor, device or controller is deployed and starts functioning, there is no room for any modification, and subsequently any changes in the device will be traced in the blockchain.

3.8 Blockchain for IoT Security and Privacy: The Case Study of a Smart Home

The major objectives of the smart home are safety, energy economy, reduction of operational expenses and expediency. Making a home smart reduces the stress on the residents, saves time and money and avoids the wastage of energy. It improves the lifestyles of the residents at home.

We have the overview of the smart home setup, which comprises various devices, processes and mechanisms connected to the Internet (Figure 3.2). Since we have seen the large-scale IoT systems in the previous topic which includes all the necessary cutting-edge technologies, we can view a smart home too with those technologies. In the proposed case study, let us have IoT and cloud computing as major building blocks integrated to experience the vital power of blockchain.

Figure 3.2 Blockchain-based smarthome system. (From Masak, K. 2018. VIONEX ICO: Blockchain-based smarthome system. *Coin Spectator*. Available at: https://coinspectator.com/news/238572/vionex-ico-blockchain-based-smarthome-system.)

IoT of smart homes encompasses home appliances such as refrigerators, air-conditioning, washers, fans, lights, televisions along with internet connection and mobile appliances management. Various sensors are incorporated with different devices, and data communication means a lot for future processing and analyzing. This adds intelligence to the devices at home to quantify the home environments and the status of the functions of the appliances.

Cloud computing can be devised to deliver effective use of all its services with respect to power of computation, data storage and applications for constructing, sustaining and running services at home and gaining access to IoT devices at home from anywhere at any time.

Smart Home Manager (SHM) acts as a central unit and works as a blockchain, handling all the transactions. It takes all inward- and outward-bound communications and makes use of a public key for local transactions with IoT devices and native storage.

The native information ledger preserves a policy header defined by the home proprietor to approve the established transactions. Overlay nodes or the devices inside the home might produce transactions with the purpose of sharing, requesting or storing data. We can assume that the cluster head which is the head of a set of devices, a node in the intersection or the cloud storage can be the adversary. The adversary is capable of fabricating transactions or deleting the data or making unwanted links to the nodes or authenticating bogus transactions.

The common network attacks which may focus on the cluster head may be a denial-of-service (DOS) attack, modification attack, dropping attack or an appending attack which will threaten accessibility for the genuine user.

The study can have a local blockchain connecting the policy header and devices using a comprehensive Diffie–Hellman algorithm with a shared key. The shared key should be assigned by the miner to devices for direct communication with each other to accomplish user control over smart home transactions. Keeping data on the local storage, each device is required to be authenticated to the storage using a shared key. The locally stored data can be moved to the cloud storage in what is known as a store transaction which is an anonymous process. The additional likely transactions are access and monitor transactions. These transactions are primarily produced by the home proprietor to observe the home when she/he is outside.

Nevertheless, the adversaries are not capable of breaking the encryption. There are major threats such as accessibility threats, anonymity threats and authentication and access control threats.

These threats can prevent the genuine user from having access to the data or services or they can find the identity of the user to breach the privacy or try to make the adversary the genuine user.

In the first level of defense, Smart Home Manager finds any data packets that violate the policies, then these packets will be dropped. The second-level defense is that any device attached to the local blockchain will be allowed to make a transaction only after genuine authentication at SHM. If any device is found to be genuine it will be isolated from the network.

Chapter 4

Consensus Algorithms – A Survey

R. Indrakumari, T. Poongodi, Kavita Saini, and B. Balamurugan

Contents

4.1 Introduction

Blockchain is considered one of the technologies with the most potential [1]. Bitcoin, proposed by Nakamoto [2], attracted researchers and industrialist attention towards blockchain as it has the capacity to eradicate the limitations of the traditional payment method which depends on a third party. In conventional payment methods, while making a payment people trust a third party who verifies the validity of their transactions. In most cases, the third party is not trustworthy as every transaction is based on a single organization, causing insufficient trust. This can be addressed by using many independent organizations, which changes the view from centralization to decentralization. Satoshi introduced the ledger design that is called block, and contains a verified transaction. Genesis block [3], considered as the first block, contains the first transactions of Bitcoin.

When a transaction occurs its validity is verified by some nodes. Validity here refers to the sufficiency of money with the sender and the digital signature of the sender [4]. After verifying the validity, the block holding the transaction is added to the chain which can be identified by all other nodes. A node can append a block holding various transactions by distributing it to other node which requested to add this node to the current chain. The limitation in this method is that, if every node requests its preferable node, then there exists a mess in the situation. To avoid this, the consensus algorithm is introduced, which holds an agreement made between all nodes about which blocks should be appended, and which nodes are permitted to append their proposed blocks. Many versions of the consensus algorithm have been proposed to date.

In this chapter, various variants of the consensus algorithm in blockchain are discussed with two main types. Initially, the proof-based consensus algorithm is discussed followed by the voting-based consensus algorithm.

4.2 Consensus

The consensus algorithm is considered as a decision-making procedure for a group in that individuals participate actively to make and support decisions that fit well for the rest. In other words it can be thought of as a resolution where the individuals are supporting the decision. The consensus algorithm is an active research topic around the world that updates the distributed shared state in a secure manner. In a traditional distributed system, fault tolerance is achieved by distributing the shared state across multiple replicas in the network. Based upon the preset state transition protocol framed by the state machine the updating of replicated shared state occurs, which is referred to as state machine replication. The concept behind replication is if one or more node crashes, it will not lose anything. The main task of the state machine is to make sure that nodes with the same inputs will produce the same outputs. These replicas contact each other to construct consensus and consent the finality of the state after a state change is executed. In blockchain-based applications, the shared state is the blockchain. Consensus can be implemented through various ways such as lottery-based algorithms like proof of work (PoW) and proof of elapsed time (PoET) or by voting-based methods which include Paxos and Redundant Byzantine Fault Tolerance (RBFT). These methods depend upon various fault-tolerance models and network requirements.

In lottery-based algorithms, the winner can be scaled to a huge number of nodes as they recommend a block and send it to the remaining nodes of the network for validation. When two or more winners propose a block, the forking method is invoked to analyze which results in a longer time to finality. In voting-based algorithms, the result is based on low latency finality. Here, the nodes transfer the message to other nodes and hence take more time to attain consensus, resulting in a trade-off between speed and scalability.

4.3 Lottery-Based Algorithms

The lottery-based algorithm is otherwise known as the Nakamoto consensus, after the founder of Bitcoin. Here a validator is elected, to make decision about the next node to be appended. The lottery-based algorithm is not an equiprobable distribution technique, as it has its own probability distribution to the winner. Various algorithms based on the lottery-based method are discussed as follows.

4.3.1 Proof of Work

Proof of work (PoW) [5] is the initial consensus protocol used for cryptocurrency that permits the blockchain users to obtain consensus in Bitcoin. This protocol particularly involves the SHA-256 hashing algorithm, Merkle tree and peer-to-peer (P2P) network to create, broadcast and verify blocks in the blockchain network. PoW also incurs costly digital computation since it includes various techniques to complete the process. The properties of PoW are described below:

- PoW is developed for permission-less public distributed ledgers and for mining processes; it consumes more computational resources.
- To construct a new block, a cryptographic puzzle must be solved by the miner, and the user who solves the puzzle first will avail the reward by broadcasting the result in the network.
- The protocol maintains the transactions in each block in a linear fashion and a block consists of the set of transactions.
- The cryptographically signed transaction will be accepted only if the signature is validated and verified in the network.
- The challenge-response computation process is known as mining.

- A reward is distributed fairly in this protocol. If the miner is determined with 'p' fraction of whole computation power, they would have a probability 'p' to mine the subsequent blocks.
- If any conflict arises, then the protocol releases multiple branches of blocks; however the longer one is retained as the trusted branch.
- The main objective of PoW lies in managing the consensus; a newly entered node can spot the current state of the network based on the protocol rules [6].

PoW introduces mining that involves a step for validating a block (group of transactions) in the network by displaying the computational proof of the completed work. Once a transaction is started, the available miners in the network compete each other to become first by solving a cryptographic puzzle and form the block. The miner who solves the puzzle successfully broadcasts the solution to the block among other peers, and the solution is verified to make the new block acceptable on the chain. Some of the implementation details are described below:

a) Bitcoin: It is the first P2P cryptocurrency that permits two participants to exchange their payments without any third-party intervention. From the beginning, it inspired many sectors such as healthcare, governance, the financial sector, etc. The payments in Bitcoin are exchanged with identity anonymity and trivial fees. As it is decentralized, it avoids the risk of counterparties and is not influenced by any of the policies of financial organizations. Micropayment channels are allowed in Bitcoin via the native protocol library [7] and offline lightening network [8]. The computational data can also be sold through zero knowledge proof to obtain the maximum trust during a transaction [9]. A multi-signature transaction is also supported to improve the security level [10, 11].
b) Litecoin [12] is an open-source P2P cryptocurrency implementation based on proof of work. It utilizes enhanced security algorithms that are both memory- and computationally intensive. Scyrpt is used to prevent counterfeiting in the consensus protocol.
c) Other cryptocurrencies implementing PoW are Primecoin, Zcash, Monero, Vertcoin, etc.

Analysis:

- PoW is a power-consuming protocol that needs a huge amount of computation power which is merely a wastage of resources as many efficient protocols are available.
- In PoW, the difficulty level also increases and with it the power required to solve the cryptographic puzzle. Moreover, it becomes inaccessible for solo miners to actively participate in the network.

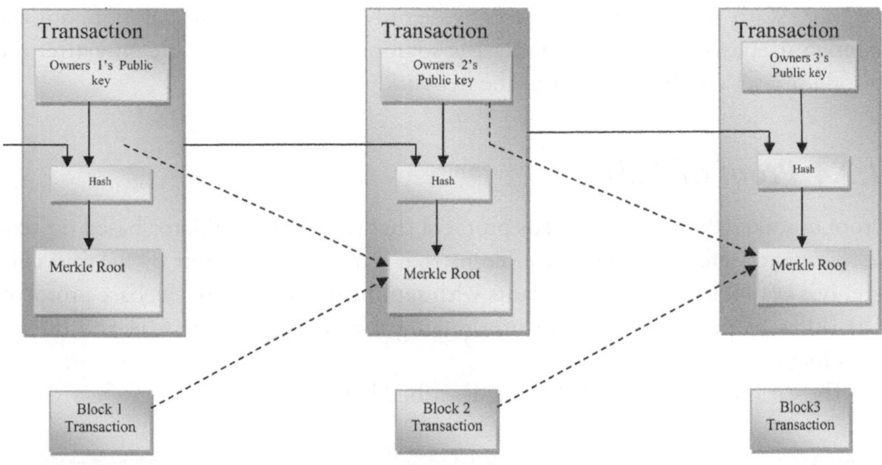

Figure 4.1 Structure of proof of work.

- The protocol is perceived as wasting immense resources due to the extensive power consumption. Other consensus protocols are preferably recommended for better output and efficient processing.
- The requirement of high computational power by PoW also guarantees a high level of security. An attacker requires 51% of computational power which is simply impossible considering the difficulty level of this protocol. However, the PoW is highly vulnerable to Sybil attacks, denial-of-service (DoS) attacks and selfish mining attacks.
- An application-specific integrated circuit (ASIC) is the hardware that manages the mining process, and it offers an unfair edge when compared to others in the network due to its expense (Figure 4.1).

4.3.2 Proof of eXercise (PoX)

It is a conceptual consensus protocol designed for the distributed ledger which consumes huge computational resources of the system. An attempt is being made to transform the hash-based PoW puzzle-mining process into a useful output in order to avoid wastage of resources. A variant of PoW solves real-world computational problems by considering matrices as an exercise. Several matrix-based real-world scientific problems in various applications include DNA-RNA matching, image processing, data mining, etc.

4.3.3 Proof of Useful Work

The conceptual idea was proposed to solve scientific problems focusing on orthogonal vectors (OV) as the proof of work, and it also integrates zero-knowledge proof concept. With this, the miners can provide only the proof of the solution, not the solution itself, to the delegated task. The solution becomes available only after the certain pre-set condition is met in the network.

4.3.4 Proof of Stake

Proof of stake (PoS) is a consensus protocol that selects the validator based on economic stake (refers the amount of coins that particular validator owns) and coin age. It is available in so many variants with significant changes in the base protocol. The different protocols differ in minimizing the centralization issue and double spending.

The various properties of PoS are described below:

- The computational challenge-response process in the protocol is known as minting.
- Initially, it was designed for permission-based public distributed ledgers and focuses on economical based puzzle solutions.
- The new coins are not being generated in PoS, hence there is no block reward and only the transaction fee will be taken by the miner in PoS.
- A new node always requires rules, protocol messages and recent state to reach the current state of the blockchain network.
- The miner for a specific block is chosen based on the economic stake in the network.
- In PoS, the probability 'p' of a validator is directly proportional to the fraction 'p' of the stake that the miner owns out of all in the round.

In PoS, the distributed ledger keeps track of the validators with their respective stake in the network. The validators in PoS invest stake to gain chances to mine the next block. The chances are higher for the validator who has the higher stake. The validators will be chosen randomly for block creation. For any cheating attempt, the stake will get debited in the system. Moreover, the block creation process in PoW is straightforward, and computational power is not significantly required.

Ethereum [13] is an open-source blockchain influenced by PoS to reach the consensus. Initially, it was based on PoW cryptocurrency; later the consensus mechanism was shifted to Proof of Stake, and it became more secure and energy-efficient. Smart contract is available to perform the operation in the blockchain network. The Ethereum platform provides a blockchain development stack, in which the developers can construct and deploy distributed apps (DApps). Huge opportunities are available to form unlimited ideas by using this promising technology in the blockchain. Other PoS-based cryptocurrencies are Peercoin, Navvcoin, Neo, Decred, Dash, PivX and Reddcoin.

Analysis:

- PoS is energy-efficient and profitable for many stakeholders.
- It is an eco-friendly protocol since it requires a negligible amount of computation. In addition, it does not need any specialized hardware for active participation.
- In PoS, more than 50 percent of power is required by an attacker to corrupt the network, and it is easier when compared to obtaining 51% in PoW. To prevent such security attacks, an economic penalty approach is followed in PoS to penalize the colluding participant. In fact, it is very effective because only major stakeholders can influence the network, and they will try to avoid penalties in the network. The penalty scheme is successfully implemented in the Ethereum platform; others have followed different strategies in order to solve this problem.

4.3.5 Delegated Proof of Stake (DPoS)

Delegated proof of stake (DPoS): It is viewed as the common variation of PoS, where the validators are elected by the stakeholders rather being validated by themselves. DPoS works based on representative democracy whereas PoS follows the direct democracy. The person who is holding wallet can vote for the validator in order to create a new block. Validators can be combined with each other to create a new block instead of competing with each other as in PoS and PoW. It encourages better opportunities for the distribution of reward as voting for a normal delegate, who in turn will give rewards back to them, thus results in decentralization. The voters should ensure the honest attitude of the validator in order to ensure the guarantee of the stake. BitShares and Steem are the most popular implementations of DPoS.

4.3.6 Leased Proof of Stake (LPoS)

It is the least commonly used variant of PoS focusing on the 'rich get richer' issue. It motivates the participants to lease the stake to vote for the node and the new block would be created by the node which has more stake. Then the received reward will be distributed amongst all leasing participants. The system also motivates the number of leasing participants to achieve rewards, hence improving the protocol's security.

Use-cases: The technology is best suited for developing a public transaction system. It is more secure and efficient for the construction of public cryptocurrencies.

4.3.7 Proof of Elapsed Time

Proof of elapsed time (PoET) [14] is an efficient consensus protocol that influences the utilization of a trusted execution environment (TEE). It extends proof of ownership and proof of time in order to enhance the efficiency of the mining process by incorporating a fair lottery system. Random waiting time is enforced for block creation by leveraging the capabilities of TEE. PoET uses Intel-based hardware (e.g. Intel SGX), and it is specifically designed for permission-less public distributed ledgers. The participants' and transaction logs are transparent and verifiable, showing more reliability of the network.

The systematic procedure of this protocol is similar to PoW, but it consumes less computational resources. The nodes compete among themselves in order to solve a cryptographic puzzle and search the next block. In the PoET protocol, each validator is assigned 'T' a random wait time to construct the block and it is tracked by it. The validator who has successfully completed the waiting time can create and publish the block in the network. The protocol follows both first come first serve (FCFS) and a random lottery scheme. The entire process relies on Software Guard Extensions (SGX) which assure trusted code execution in a safe environment (i.e. Intel Software Guard Extension).

PoET reaches consensus by maintaining the anonymity of the network participants. A monotonic counter-type hardware is maintained in TEE to protect the system from malicious activities which also ensures that only one instance is currently executing in a single CPU. There may be the chance of creating multiple instances of 'T' wait time by the participants in order to boost their luck. The protocol is highly susceptible to various security attacks and lacks security analysis [10]. Particularly, Intel Software Guard Extension is vulnerable to rollback attacks [15].

Hyperledger Sawtooth [16] is a modular blockchain introduced by Intel, and it follows the PoET consensus algorithm for implementing a leader election lottery system. Parallel processing is followed in transactions for block creation and validation by using 'Advance Transaction Execution Engine'. The protocol is highly capable in providing efficient throughput among a huge network population. Moreover, it is an enterprise grade protocol that enables the development process of general-purpose smart contracts.

4.3.8 Proof of Luck (PoL)

Proof of luck (PoL) is a conceptual permissioned consensus protocol based on trusted execution environments (TEEs) (i.e. Intel SGX) [17]. It extends the functionality of proof of ownership and proof of time, addressing the issues such as centralization of available consensus protocol (PoS, PoW) and extensive energy consumption. Moreover, it exhibits low latency of transaction validation, and the block confirmation time is 15 seconds greater when compared to Ethereum and significantly 10 minutes less than Bitcoin.

The protocol signals the participants in each round to commit all the available uncommitted transactions to a new block and the version block is assigned a numeric value. Later, the voting process is started in which the participants vote for a number randomly and the node with the highest vote wins the luckiest block. The other participants in the network stop the mining process and their own block is broadcasted as soon as the luckiest block is received; hence the network congestion can be minimized.

4.3.9 Proof of Space or Proof of Storage

Proof of space or proof of storage [18] is an ecofriendly protocol developed to avoid abuse of resources [19]. It is similar to proof of work but instead of computation it involves disk consumption.

- Proof of space is meant for public distributed ledgers, and the free disk storage is considered as the resource.
- The influence of a miner's power over the network is directly proportional to the amount of disk space being contributed.

4.3.9.1 Theory

Proof of storage utilizes disk space to mine a block. It verifies the honesty of a remote file by distributing a copy of data to a server and computing a challenge-response protocol to ensure the integrity of the data. The actors in the proof-of-storage algorithm are the Provers and the Verifiers. Provers are actors who store data, and Verifiers are actors who authenticate that the Provers are storing the data. Verifiers usually provide a challenge to the Provers, who in turn solve the challenge with a proof to the exact proof-of-storage scheme.

Proof of storage generates random solutions called plots in advance using the Shabal algorithm and saves it on hard drive. This process is called plotting. Following the plotting process, the miners compare the solutions with the recent puzzle [20].

4.3.9.2 Burstcoin

Burstcoin is a mineable coin implemented with an eco-friendly proof-of-space algorithm in 2014. It is a decentralized cryptocurrency and payment system that depends on space when mining resources [21]. Burstcoin mining is inexpensive, and it can be performed on a mobile device [22]. The first Turing complete smart contract which solves computing problems is implemented using proof-of-space protocol.

4.3.9.3 SpaceMint

SpaceMint is a cryptocurrency, replacing energy-intensive computations associated with cryptocurrencies by proof of space. Here miners invest disk space instead of computing power. The mining process takes place in two phases, initialization and mining. In initializing, the miners contribute N bits of space and create secret key pairs. The miners publish its space commitment through a special transaction. In the mining phase, mining is incentivized through block rewards and transaction fees. Once initialized, each miner attempts to add a block to the blockchain every time period. SpaceMint holds three types of transactions, namely, payment, space commitments and penalties. Each transaction is signed by the users and sent to miners to be added to the block.

4.4 Voting-Based Consensus

In voting-based consensus algorithms, the verifying network should be adjustable and explicitly known in order to exchange the message without any complication. In proof-based consensus algorithms, the nodes are allowed freely to unite and pullout from the network.

The nodes in the voting-based consensus algorithm communicate with each other prior to adding their own blocks into the chain. The execution process is same as the conventional fault tolerance method incorporated in the distributed system [23].

As in any fault-tolerance method, the voting-based consensuses are intended to work when there is a crash in the nodes and sometimes the nodes are subverted.

When the node crashes, it waits for the information passed by other nodes. In some cases, the waiting node will not receive any proper message or guidance from other nodes to make a decision. To prevent this, there should be n + 1 nodes instead of n nodes to perform uninterrupted operation [24].

In contrast to this, the subverting nodes perform outlandish operations, resulting in imprecise output. These issues can be addressed by a classical problem, popularly called the Byzantine Generals Problem developed by Lamport et al. [25].

Here, the concept is that Byzantine generals have occupied an enemy's camp by dividing their army force into N groups under N generals, who are capable of attacking the enemies from various sites. To win, the N groups of armies should attack at the same time. Before commencing the attack, they should come to an agreement about the time of attacking by exchanging proper messages, and the decision is taken by the majority. Regrettably, there is some conspirator among the general group whose intention is to cheat the other generals by passing diverse decision to others which causes failure in the attack as some generals are not participating in the attack.

The solution for this issue is proposed by Lamport et al. that to tolerate the n subverted generals, there should be at least 2n + 1 generals to accompany them. The same scenario is applicable in blockchain in that some nodes can be subverted when executing the consensus work by propagating diverse results to other nodes. These bad situations lead to the classification of voting-based consensus algorithms into

1. Byzantine fault tolerance-based consensus which avoids the occurrence of crashing and subverted nodes.
2. Crash fault tolerance-based consensus which prevents the cases of crashing nodes.

The consensus algorithms under these sub-categories make an assumption that among N nodes, there should be at least t nodes (t < N) operating normally. While in crash fault tolerance-based consensus, t is usually set equal to [N/2 + 1], in Byzantine fault tolerance-based consensus, t is usually assigned equal to [2N/3 + 1].

4.4.1 Byzantine Fault Tolerance-Based Consensus

Byzantine fault tolerance-based consensus is based on the popular Hyperledger Blockchain platform [26] used by many enterprises, especially IBM [27]. Castro and Liskov [28] proposed a variant of Byzantine fault tolerance called the Practical Byzantine Fault Tolerance (PBFT) intended for Hyperledger Fabric [29]. In Byzantine fault tolerance, there exist two sorts of nodes, namely, a leader node and validating peers.

Formerly, the validating peers receive the request from the clients to validate the transactions. After validating, the results are sent to the leader and other peers. Here the threshold is the batch size is maintained. Based on the time of creation, the leader arranges the transactions and puts them into a block.

Symbiont [30] and R3 Corda [31] are famous blockchain platforms based on Byzantine fault-tolerance consensus algorithms proposed by Bessani et al. [32]. In addition to the procedure of execution, Bessani et al. developed a replica for storing the log of executed operations in a single machine, which is used for gaining the last current state, when a node fails, and needs to restart.

4.4.2 Crash Fault Tolerance-Based Consensus

Paxos [33] and Raft [34] is a crash fault tolerance-based consensus used by Quorum [35] to tolerate crashes. Raft is based on an assumption that every time, [n/2 + 1] of the total nodes work normally. In the Raft consensus algorithm, the verifying nodes take the role of follower, candidate and leader. The communication among the nodes is made by messages: RequestVote for voting a leader node, and AppendEntries for transferring the requests to other nodes.

During execution, the requested transactions from the clients are received by the leader who in turn saves them to a list called log entry. After receiving the request, the leader sends the AppendEntries message to all followers which contains the transaction log along with previous transaction index. For instance, if the leader sends the nth transaction, then he should attach the (n – 1)th transaction details.

Chain [36], a blockchain platform, uses an algorithm called federated which is based on the crash fault-tolerance consensus algorithm. Here there are n nodes in the verifying network; among them there are two nodes called the block generator and block signers. The transactions received from the clients are verified by the block generator and the valid one is saved in a temporary list. The block generator considers some requested transactions sequentially and puts them into the blocks and circulates to all block signers. The blocks received by the signers are verified, and the valid blocks are signed and sent back to the block generator. If a block is signed by a majority of block signers, then it is considered as a trustworthy block and the same is appended to the chain maintained by the block generator. Since more than one block signer acknowledges the block, if there exists any crash then the chain could resist the crash fault.

4.5 Conclusions

This chapter provides a survey of some consensus algorithms applicable in blockchain. The algorithms are categorized into two types, namely, proof-based algorithms and vote-based algorithms. In the proof-based algorithms, the nodes have to prove its majority to append the blocks which it requires. In the vote-based algorithm, an agreement is made among nodes regarding the blocks to be appended to the ledger. The applications of these two types of algorithms are discussed elaborately in the chapter.

References

1. S. Haber and W. S. Stornetta, "How to time-stamp a digital document," *Journal of Cryptology*, vol. 3, no. 2, pp. 99–111, 1991.
2. S. Nakamoto, "Bitcoin: a peer-to-peer electronic cash system," 2008 [Online]. Available: https://bitcoin.org/ bitcoin.pdf.
3. Bitcoinwiki, "Genesis block," 2017 [Online]. Available: https://en.bitcoin.it/wiki/Genesis_block.
4. E. Robert, "Digital signatures," 2017 [Online]. Available: http://cs.stanford.edu/people/eroberts/courses/ soco/projects/public-key-cryptography/dig_sig.html.
5. S. Nakamoto, "Bitcoin: a peer-to-peer electronic cash system (white paper)," 2008. [Online]. Available: https://bitcoin.org/bitcoin.pdf.
6. R. Greenfield, "Vulnerability: proof of work vs. proof of stake," 27/08/2017. [Online]. Available: https://medium.com/@robertgreenfieldiv/vulnerabilit y-proof-of-work -vs-proof-of-stake-f0c44807d18c.
7. J. Poon and T. Dryja, "The Bitcoin lightning network: scalable off-chain instant payments," 26/01/2016. [Online]. Available: https://lightning.network/lightning-netw orkpaper.pdf.
8. Bitcoinj Community, "Working with micropayment channels," [Online]. Available: https://bitcoinj.github.io/.

9. Gmaxwell, "Zero knowledge contingent payment," 02/2016. [Online]. Available: https://en.bitcoin.it/wiki/Zero_Knowledge_Contingen t_Payment.
10. M. Rosenfeld, "What are multi-signature transactions?" 18/05/2012. [Online]. Available: https://bitcoin.stackexchange.com/questions/3718/wh at-are-multi-signature-transactions.
11. Belcher, "Multisignature," 12/2018. [Online]. Available: https://en.bitcoin.it/wiki/Multisignature.
12. Litecoin Project Community, "About LiteCoin," 2018. [Online]. Available: https://litecoin.org/.
13. J. Ray, "Ethereum (Whitepaper)," 26/05/2018. [Online]. Available: https://github.com/ethereum/wiki/wiki/White-Paper.
14. L. Chen, L. Xu, N. Shah, W. Shi, Z. Gao and Y. Lu, "Security analysis of Proof-of-Elapsed-Time (PoET)," In *SSS 2017*, Boston, MA, 2017.
15. M. Brandenburger, C. Cachin, M. Lorenz and R. Kapitza, "Rollback and forking detection for trusted execution environments using lightweight collective memory," In *Conference: 2017 47th Annual IEEE/IFIP International Conference on Dependable Systems and Networks (DSN)*, 2017.
16. K. Olson, M. Bowman, J. Mitchell, S. Amundson, D. Middleton and C. Montgomery, "Hyperledger Sawtooth (whitepaper)," 01/2018. [Online]. Available: https://www.hyp erledger.org/wpcontent/uploads/2018/01/Hyperledger_Sawtooth_Wh itePaper.pdf.
17. M. Milutinovic, W. He, H. Wu and M. Kanwal, "Proof of luck: an efficient blockchain consensus protocol," In *Middleware Conference*, Italy, 2016.
18. S. Dziembowski, S. Faust, V. Kolmogorov and K. Pietrzak, "Proof of space," In *International Association for Cryptologic Research (IACR)*, 2013.
19. gmaxwell, "Proof of Storage to make distributed resource consumption costly," 10/2013. [Online]. Available: https://bitcointalk.org/index.php?topic=310323.0.
20. P. Andrew, "What is proof of capacity? An eco-friendly mining solution," 31/01/2018. [Online]. Available: https://coincentral.com/what-is-proof-ofcapacity/.
21. S. Gauld, F. V. Ancoina and R. Stadler, "The burst Dymaxion," 27/12/2017. [Online]. Available: https://www.burst-coin.org/wpcontent/uploads/2017/07/The-Burst-Dymaxion-1.00.pdf.
22. P. Andrew, "What is Burstcoin?" 31/01/2018. [Online]. Available: https://coincentral.com/what-isburstcoin-beginners-guide/.
23. W. L. Heimerdinger and C. B. Weinstock, "A conceptual framework for system fault tolerance," Defense Technical Information Center, Technical Report CMU/SEI-92-TR-033, 1992.
24. L. Lamport, "Paxos made simple," *ACM SIGACT News*, vol. 32, no. 4, pp. 18–25, 2014.
25. L. Lamport, R. Shostak and M, Pease, "The Byzantine generals problem," *ACM Transactions on Programming Languages and Systems*, vol. 4, no. 3, pp. 382–401, 1982.
26. Hyperledger [Online]. Available: http://hyperledger.org/.
27. Hyperledger fabric [Online]. Available: https://github.com/hyperledger/fabric.
28. M. Castro and B. Liskov, "Practical Byzantine fault tolerance," In *Proceedings of the Third Symposium on Operating Systems Design and Implementation*, New Orleans, LA, 1999, pp. 173–186.
29. C. Cachin, "Architecture of the hyperledger blockchain fabric," In *Proceedings of ACM Workshop on Distributed Cryptocurrencies and Consensus Ledgers*, Chicago, IL, 2016.

30. Symbiont [Online]. Available: https://symbiont.io/.
31. Corda [Online]. Available: https://www.corda.net/.
32. Bessani, J. Sousaand E. E. P. Alchieri, "State machine replication for the masses with BFT-SMART," In *Proceedings of 2014 44th Annual IEEE/IFIP International Conference on Dependable Systems and Networks*, Atlanta, GA, 2014, pp. 355–362.
33. L. Lamport, "Paxos made simple," *ACM SIGACT News*, vol. 32, no. 4, pp. 18–25, 2014.
34. D. Ongaro and J. K. Ousterhout, "In search of an understandable consensus algorithm," In *Proceedings of 2014 USENIX Annual Technical Conference*, Philadelphia, PA, 2014, pp. 305–319.
35. Raft-based consensus for Ethereum/Quorum [Online]. Available: https://github.com/jpmorganchase/ quorum/blob/master/raft/doc.md.
36. Federated Consensus [Online]. Available: https://chain.com/docs/1.2/protocol/papers/federated-consensus.

Chapter 5

Optimized Digital Transformation in Government Services with Blockchain

R. Sujatha, C. Navaneethan, Rajesh Kaluri, and S. Prasanna

Contents

5.1 Overview of Blockchain

The blockchain concept relies on cryptography that links the blocks and each block holds the cryptographic hash of a previous block and timestamp along with data to be transferred. The openness in the distributed setup is achieved with the help of a peer-to-peer network. The main idea is that the need for central trusted third party is obsolete. Each node, another name for block, possesses a complete replica of the blockchain. Adding a new transaction, that is, adding a block, ensures that block is appended to all blocks such that transparency is maintained. Tampering with data is not possible, and the working of a system is flawless when integrated with blockchain technology. Cryptographic arrangements help in increasing the security of the transaction in any sort of application. The type of restriction blockchain is broadly classified into public, private and consortium. The blockchain provides reliable, immutable, irrevocable, transparent data available at all times, with reduced costs because a third party is not required.

Initially starting with the buzz word cryptocurrency, the blockchain has found applications in various domains in both the public and private sectors across the globe. A meeting conducted by the Organization for Economic Cooperation and Development (OECD) discussed the uses and limitations of blockchain in the public sector in October 2018. It mentioned clearly that the blockchain is progressing exponentially. The meeting stats collated by the Illinois Blockchain Initiative in April 2018 show 203 blockchain initiatives in 46 countries are booming. Among those initiatives, a few are in exploration, so much at the strategy stage, a few are in the prototyping and incubation state and a few are up or live. The blockchain is utilized actively in public sectors like transport management, taxation, voting, land registration, health care, identity management, digital payment, and the list goes on.

In particular payment systems which do not involve cryptocurrency also need digital identification. To promote the export–import of a country, identifying legitimate payment methods is a difficult task and involves dealing with a lot of complaints. A blockchain-based digital identity that incorporates permissive blockchain as one of the regulatory nodes that are incorporated in all transactions will be a viable solution for the government sector.

Health care is a primary concern of all developing countries since there are many registered alternate medicine practitioners and methodologies, and with millions of illiterate people, it is cumbersome to maintain and retrieve the health records of millions for future use for health care schemes. Blockchain can provide a sustainable solution for the maintenance and retrieval of health records.

Even many of the regulatory mechanisms and departments can be incorporated with a blockchain which provides a genuine authentic platform without any intermediate intervention to common people.

Blockchain technology sounds like a viable solution for many departments due to its decentralized distributed self-regulated nature. Even though there are many critics of the first, most successful system, Bitcoin, by this technology all federal governments across the globe understand the potential sustainable benefits of blockchain technology. The successive improvement blockchain 2.0 and with the Hyperledger (claimed as 3.0) have addressed the issue of scalability of the blockchain. So by incorporating blockchain technology, a sustainable, adaptable, decentralized system can be deployed which also consumes fewer resources and requires almost zero interference from the government. The permissive blockchain will also ensure that in the name of decentralization the regulator need not totally lose control. The system can be monitored, and to handle abnormalities the regulator can enforce actions using a special privilege consensus mechanism which is possible with blockchain 3.0.

5.2 Blockchain in Traffic Management

Many of the fast-growing countries, as well as developed countries, have many urban areas and metro cities. The governments of developing and developed nations are finding out better ways to manage the growing cities and taking effective measures to curb traffic congestion that leads to over-pollution. So the issue of smart mobility becomes a nightmare for almost all governments across the globe. The defined government may provide a framework and handle mobility for their public transport since they might have complete information about mobility. However, this kind of centralized system is limited to government-approved transport and mobility services only. All the urban areas across the globe have huge unaccounted highly dynamic private vehicular mobility about which centralized traffic management cannot make any prediction. This added mobility issue will make the estimates of centralized public transport systems imprecise, and hence the citizens may be frustrated with inaccurate prediction results by the so-called centralized mobility management systems. It is quite common that many nations have well-regulated public rail and road transport services, and there are several web and mobile applications running presently to address the needs of smart mobility. But the significant thing is this kind of application mostly does not take into account a sudden outburst of private unpredictable traffic which may defeat the purpose of the smart mobility features incorporated in place.

Sharma proposed a sustainable approach that inherently reduces the latency, and addresses the scalability, privacy and bandwidth issues. Their hybrid architecture has the added advantages of decentralized and centralized architectures. The privacy and security are assured in the proof-of-work scheme provided (Sharma & Park, 2018). Singh worked on a theme called intelligent vehicle (IV) which make use of blockchain technology to create a trusted environment for communication while assuring privacy. It encompasses a local dynamic blockchain that is compared with the main blockchain to examine the trustworthiness of the network. The entire branching process is automated by unbranching and branching algorithms as well (Singh & Kim, 2018). Chuvan-chi proposed a mobility-aware data approach

(MADA). This work reduces the data dissemination and also the overhead of maintaining the shared data consistency. Safe time and speed of updating are the significant parameters taken for determining the mobility. It reduces the number of messages that require retransmission and also the duplication of messages. The task of maintaining data consistency is simplified (Lai & Liu, 2019).

5.3 Blockchain for Taxation

Blockchain is an emerging technology which has gained interest from various sources like energy industries, startups, financial institutions, supply firms, national and international governments, etc. These sources are aiming, from different backgrounds such as voting, taxation and land registration, to identify blockchain, which has potential drives to bring substantial welfares and innovation. A forum from the world economy has recognized blockchain technology as one of the major trends in the forthcoming evolution of the digital world (Beck et al., 2017).

Blockchain technology aims not to have a centralized system across the connected peers so that all the parties can have easy access in a secure environment. The information that has been stored in the block need not be a currency, and it can be of any kind of data. Blockchain usually fits into all the different industries and multiple businesses sectors by which anyone can access information stored on it (Wijaya et al., 2017).

The key part of the blockchain process is achieving accord among the gatherings to add data to the existing block. This technology separates the worldview of the concentrated accord, and brought together the framework is utilized to lead on legitimacy. An accord component guarantees that the additional block contains honest data.

5.4 Blockchain Could Transform the World of Indirect Tax

A few years ago, there was no single word of blockchain related to finance and taxation. In recent years, investors and professionals have been aiming to use blockchain in all the applications of finance (Beck et al., 2017) (Pokrovskaia, 2017). Various sectors of finance such as trades, banks and exchanges are keenly looking at the growth of blockchain in finance. Blockchain has just set some pilot projects in digital payment platforms and many applications are to come in the near future. The professionals from the tax department have queries like

- Do the ledgers from the distributed systems remove the need for invoices?
- How will the government policies levy cryptocurrencies of the individual and refund and collect the various taxes?

- Can customs declarations be made automatically instead of by customs brokers and others?

  Tax policies and their problems will vary from country to country. In order for an individual to file their tax, a tax system always encounters some complicated problems. Someone who knows the tax laws of a tax system are also not able to perform accurately, since every time the regulations in taxation will be changed. With an increase in regulations, the complexity in taxation also increases rigorously (Schwanke, 2017). Taxes are collected from individuals and business people as per the profits that they made in a financial year. There are also some other specific taxes that are taken such as GST, sales tax, value-added tax (VAT).

Taxes like GST, VAT, etc. are the vital source of income (Ainsworth & Viitasaari, 2017), (Ainsworth & Shact, 2016) for local administrative organizations and change contingent upon the item or administration they are exacted on, just as the region of the locale they are connected to. These kinds of taxes are difficult to monitor both by administrative organizations and the general population who need to dispatch these charges. Now and again a huge segment of these expenses goes unpaid, denying state offices some truly necessary financing (Budish, 2018). Because of the complications and indeterminacies in the assessment and legitimate frameworks joined with the failure of state offices to demonstrate and follow these acts of neglect, various high-acquiring elements frequently pull off covering absurdly low regulatory expenses, while many pay a lot.

Blockchain is very much essential in the field of taxation so that many changes can easily happen and the efficiency of the taxation system will increase and thereby every individual tax account will be safeguarded. Figure 5.1 illustrates a working model of blockchain under the taxation system. Blockchain taxation can be revolutionary by making the following promises:

- Nobody can modify or disturb the committed blocks of the respective blockchain system.

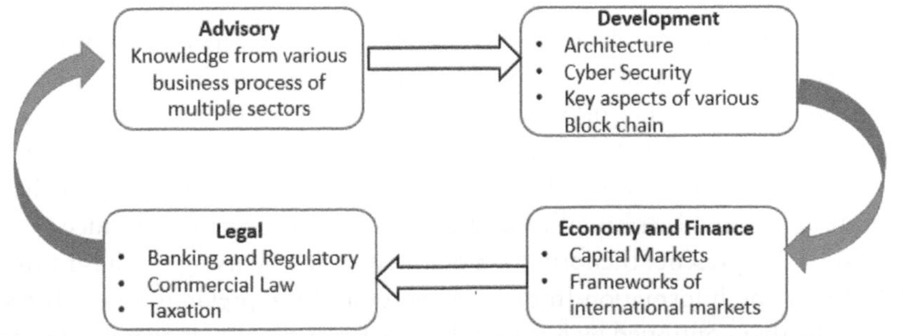

Figure 5.1 Working model of blockchain under the taxation system.

- Immutability creates the origin of the money and thus makes tax calculation easy.
- By checking the capital and ownership of the assets owned, the network systems can be transparently enabled.

5.5 Blockchain Technology for Voting

In the course of the last 25 years, cryptographers have created decision conventions that guarantee an extreme change in the perspective on election results which are to be checked completely by the spectators (Adida, 2008). Voting in favor of anything is a standout amongst the most critical approaches to guarantee reasonable portrayal and equivalent voice; it can end up hard to precisely and effectively monitor every voter's qualification and authenticity to take an interest. Over this, there are likewise different issues, including absence of straightforwardness and potential debasement which debilitate many from casting a ballot by any means.

An e-voting framework must have increased security all together to ensure it is accessible to voters. However it must be secured against external impacts like altering votes, and must protect citizens' votes from being messed with. Numerous automated ballot-casting frameworks depend on Tor to conceal the identities of voters (Ayed, 2017). Usually, the voting process is decentralized; this means that there is no proper trusted agency to conduct the elections online fairly. In the recent past, many systems of voting have been centralized schemes. Customary databases are kept up by a solitary association, and that association has unlimited oversight of the database, including the capacity to alter the stored information, to blue pencil generally legitimate changes to the information or to include information deceitfully.

In recent years, many traditional voting systems have been used between the voters and ballots, along with the public-key cryptographic elements and Blind Signature Theorem (Chaum, 1981). According to the Estonian I-Voting System (TRUEB, 2013), voting can be done completely by using the Internet along with a valid government identity card. In this process, if the voter tries to cast a vote several times then the last vote cast is considered. Finally, the data will be stored in the election servers.

As of late, the dispersed electronic casting of a ballot dependent on the blockchain is the potential area to explore (McCorry et al., 2017). Some blockchain-based ballot-casting frameworks are prevailing now. Be that as it may, the majority of them simply utilize blockchain as capacity means for casting ballot information, then apply little scale casting a ballot, followed by open key location in the first Bitcoin program and is essentially utilized in the client security insurance.

Figure 5.2 represents the formation of a new block by using the hash function HAS-256 for the information from a new voter and the previous voter. The same process is further continued by forming the other new blocks along with the crypto chain (Figure 5.3).

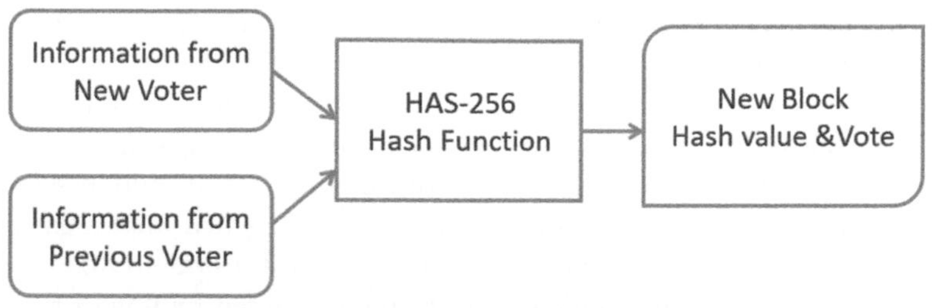

Figure 5.2 Formation of a new block for the voting system.

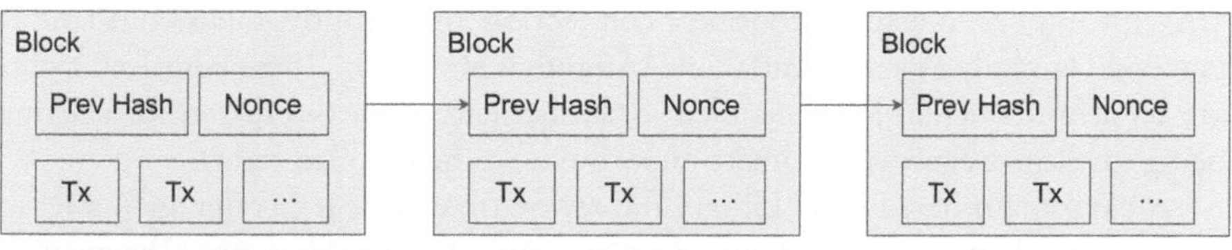

Figure 5.3 Series of blocks.

The primary exchange added to the square will be an exceptional exchange that speaks to the voter (Evans & Paul, 2004). Once the exchange begins, incorporating applicants' names as the first blocks each proceeds vote in favor of the particular competitor. In contrast to different exchanges, the establishment won't consider a vote. It will just contain the label of the applicant (Noizat, 2015) (Wright & De Filippi, 2015). With respect to the e-Casting, a ballot framework will permit a dissenting vote, where the voter may return a clear poll to show disappointment with all options or refusal of the existing political framework as well as the race. Each time an individual votes the exchange will be logged and the blockchain will be restored.

At the outset, our general public is improving and propelling, and it's basic that we exploit our innovation to guarantee that our casting ballot forms are as secure as could reasonably be expected (Wang et al., 2018). The most ideal approach is to investigate the potential employment of blockchain to improve the decision procedure in general. This disposes of the need for any expert to check and legitimize the votes (Hanifatunnisa & Rahardjo, 2017). Rather, the disseminated record innovation does that consequently, guaranteeing that each vote is genuine and affirmed. It might require investment for blockchain to be comprehended and utilized by everybody, but it's something to anticipate as our general public keeps on developing and advancing.

5.6 Blockchain for Land Registry

In various blockchain applications, land registry and settlement is a common thing in public facilities, in which the land data, for example, the physical status and associated rights, can be enrolled then announced via blockchain. Furthermore, every progression made on the land, for example, the exchange of land or else the foundation of a home loan, can be documented and overseen on blockchain, thus increasing the effectiveness of open administrations (Zheng et al., 2016). "Land Administration is the process of determining, recording, and disseminating of information about ownership, value, and use of land when implementing land management policies" (UN Economic Commission for Europe, 1996).

On the off chance that proprietorship is comprehended by way of the instrument through which rights to arrive are held, we can likewise talk about land residency. Land residency mirrors a social relationship with respect to privileges to arrive; it implies in a specific locale a connection among individuals, and land is perceived lawfully as a substantial one (Anand et al., 2016). These perceived rights are on a fundamental level qualified for enlistment, with the reason for existing being to relegate specific legitimate importance to the enrolled right.

A blockchain-based Land Library framework may appear to offer an answer to these issues, despite the fact that as a general rule it maybe does not. The genuine test for these nations will most likely be the underlying recognizable proof of right holders and the production of real titles. When it is realized who is the genuine proprietor of a specific package, the responsibility for a bundle can be exchanged (Vos et al., 2017) (Themistocleous, 2018). This underlying stage won't be acknowledged by utilizing blockchain. Blockchain is planned as a mutual single wellspring of trust, to reject (questioned) administrative gatherings and banks. However, it requires a vacant stage which everybody can concede to as a beginning stage. This stage will be placed in the primary period of a blockchain, the beginning square. This beginning stage might be the issue on account of these nations, in light of the fact that there is no trust, thus there will be no assent by every single invested individual. In those cases, a blockchain-based Land Library won't work (Swan, 2017). Figure 5.4 shows the framework of blockchain for land registry.

The point of the blockchain is to guarantee the advanced exchange of significant worth starting with one gathering, then onto the next. One of the issues of advanced data is that they tend to be replicated: when somebody has a music record and sends it to someone else, the two people have a duplicate of the music document (McMurren et al., 2018). For the sharing of data, this works fine and dandy, despite the fact that there is no permit and the craftsman does not get any reimbursement (Thomas & Huang, 2017). With regards to the exchange of significant worth, it can't be duplicated (effectively). When somebody gives a specific measure of cash to someone else, it isn't sufficient for this individual to get a duplicate of the cash; the responsibility for the cash must be exchanged. The conventional method to fathom this is to host a confided in third gathering, which monitors the cash: for the most part a bank. The blockchain takes care of the issue of twofold spending by keeping a changeless record of all exchanges that is accessible to all members in the system (Ramya et al., 2018).

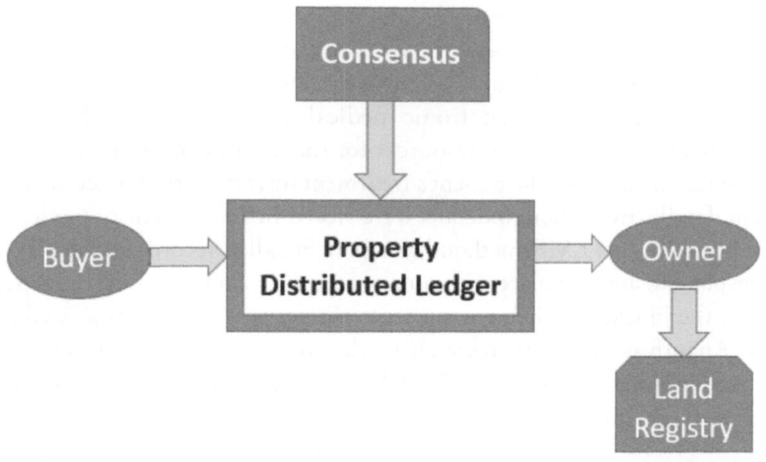

Figure 5.4 Framework of blockchain for land registry.

The blockchain creates trust by decentralizing the handling of exchanges and making this procedure straightforward. In our present society, we sort out trust by making confided in gatherings that we trust to process exchanges and actualizing governing rules by other confided in gatherings (Vos, 2016). These are on a very basic level of diverse methods for arranging trust in a framework. Note that in many frameworks, there is more trust than the blockchain gives. For instance, Bitcoin has no focal expert dealing with the estimation of the cash, while for most national bank-supported monetary standards this is the situation.

5.7 Blockchain in Health Care

Nowadays people understand that keeping data secure is indispensable. The blockchain helps us to keep data secure by allowing health record data to be moved whenever we need without any illegal activities like forgery, theft and malice. The blockchain can store data block by block; those data later linked together form a "Chain", like a food chain. The blockchain works based on a peer-to-peer network, cascaded encryption, distributed database, transparency with pseudonymity and irreversibility of records. Blockchain technology in health care involves various applications, particularly keeping patient records which can be accessed whenever needed by management. Here the blockchain ensures that even the patient is unable to access their records. Besides security and interoperability of data, the blockchain can also provide services at low costs; the devices used by hospitals are all updated on blockchain and made available whenever data may be lost. This kind of action will be taken by blockchain as per the health care concern. Even though we can share this data anywhere in the world, confidentiality is maintained by the blockchain without the loss of patient data. Moreover, the blockchain speeds up the resource and development process when any new drugs come on the market. Thus, the blockchain may make disruptive changes in the future (Sadiku et al., 2018).

Making and maintaining electronic medical records (EMR) in the blockchain is one of the precious and major resources for the health care system. Basically, an EMR provides details like the patient's treatment history, personal details and their physician details. Even though details were stored in EMR security mode, control data were stolen. This EMR method has a benefit called records and can exchange from one place to another place without any security guarantee. But using exchange records in the blockchain helps to preserve the privacy of systems as well as high security. Ananth et al., in his research work, proposed a method to keep data in secure and immutable, called the Dual-Tree Complex wavelet transform method (Ananth et al., 2018).

1. Primary care – this method arranges details about physician and diagnostic techniques used to collect the patient health status, such as symptoms of any diseases, any other problem faced by patient, and previous history of medical details.
2. Cross-disciplinary referral – the main purpose of this method is so that the hospital chief or physicians or any other health care providers can communicate about the patient's health care.
3. A multidisciplinary approach – this method works with different people to achieve a set of goals to cure the patient of the illness.

An electronic health record system may be used within the organization that was created and maintained by a particular organization. But this method doesn't suit for those who are traveling or looking for a new hospital. Such issues will happen if the patient details don't fall under their seeking hospital, which means that patients get disappointed and are inconvenienced by having to seek a hospital. To overcome this issue, we are supposed to follow blockchain advancement technology. This advancement technology is comprised of three operations, namely information confidentiality, accuracy and ready-to-use information. In this advancement of the blockchain technology method, when the user changes any information the members will verify the accuracy of the information with a code. Once data are stored in the records, the patient can neither delete the data nor edit the data in the records.

Here two methodologies are used to store and manipulate data in the EMR system.

1. Research instrument: Sample business model of the consortium blockchain type used. In this method, if any new patients join an EMR system they can allow the members access to records over the network.
2. Selection of respondents: To identify patients and physicians, respond to that person's questionnaire.

> To analyze the relationship between patients and the blockchain acceptance-based model, a multiple regression method is used. This kind of acceptance model has been used to apply the EMR system through the hospital (Wanitcharakkhakul & Rotchanakitumnuai, 2017).

Data privacy in health care in terms of keeping data from being stolen by someone is difficult, especially cognitive security in blockchain technology based on the environments called "smart ambient assisted living." Using this we can secure patient data, and only authorized parties can access this kind of data. In this process, the individual care process helps to collect patients' previous history and present health care situations. Three methods used are cognitive security impact evaluation, blockchain data privacy and protection, and personal rights and information protection. Several models and considerations are taken to implement the blockchain technology in clinical health care. Hyperledger Fabric DLA method is used for obtaining security in the environment. Top software companies like Google and Microsoft and many conferences conducted by the IEEE have been involved in ensuring security in clinical health care. Blockchain technology has achieved unsurpassed security in keeping records in the health care system.

To improve the health care system in terms of decreasing the cost and complexity, we are supposed to follow blockchain technology insurance companies. The Estonian government has taken a step to improve health care in the commercial sector and the government sector. The Estonian government created an innovation strategy that implements the blockchain technology in the whole country. They demonstrated how the process works in govtech partnerships. The population of the country year by year increases and places greater demands on the health care system. Not only is demand rising but the cost of the medicines is also increasing. In order to solve this problem, Estonia introduced an innovative plan in 2011 with a technology called blockchain in govtech partnerships. In this method a proprietary keyless signature was introduced in blockchain to ensure the security of the records, and it authorizes the availability of parties also. This blockchain-based health care records system has some benefits like its scalable, secure record system. Because of its distributed nature blockchain can easily share data only with authorized parties. The data auditing method was improved by the blockchain because blockchain records are immutable.

Using big data in blockchain technology aims to make information portable, or if any third party person requires big data, it will set permission to access those third parties. Nowadays medical industries work based on value-based business; this will help to prevent the diseases, change each and every person's lifestyle, identify infections, etc. In order to manage the complexity of health care sector data, analytics tools help to smooth the process as well as improve the efficiency of the medical practice and make workflow accurate. Here blockchain technology starts with one trusted ecosystem especially for decision making, and this blockchain involves a timestamp that helps to authenticate dataset changes; this ability provides the permission to manage the electronic health record data for one user or more than user if they edit any document purpose. Previously patients were not able to share their personal data but now they can share their data securely with even new members because here the blockchain technology used as authentication with a timestamp that makes more secure about patient record details. In this type of approach if any patient's data are lost or crash in the sense automatically bulk of data that was crashed, those data will be updated soon as well as if any of the attackers attack those data at last the attacker may get a message called failure or rejection because this whole system conditions verified by multiple system conditions. Even though we are using the current technology to enable or keep data very secure or maintain privacy a little bit of difficulty. When we use big data technology that helps to keep data security also sometimes we can break personal data. For such issues, blockchain is one possible best fit model. When we divide our data into so many categories big data with blockchain provide security to the users as well as to hospital management (Shilpa et al., 2018). In the past all records were created in handwritten format only. But now a new trend has evolved in health care to generate reports as well as data in the format of the digital chart. This new method of electronic medical records and electronic health records replaces the old method anticipated paper chart creation and cumbersome. To keep electronic records secure, one protocol helps which is based on the HIPAA Security rule. This protocol protects health records from virus attacks, making records confidential. There are two different methods used to secure data in blockchain; one is the Mooti model, and the other model is the Enigma model. The Mooti model in blockchain keeps data secure and gives access only to those who are authenticated users; the Enigma model organizes entire medical health records which include private information across a distributed network, and only the party owner can decrypt (Daniel et al., 2017).

Blockchain technology in health insurance may provide a new solution to the health care sector. This method is based on a framework that gives an efficient and fraud-free solution to the insurance claims. This framework is designed based on permissionless blockchain, an open source called "Ethereum". This process of authentication depends on the involvement of various parties over the network. Involving blockchain in health insurance decreases the amount of time and cost required for processing. There are technical challenges. In order to overcome these, blockchain cannot solve the data standardization problem but can provide a data share method in real time only in the trusted network. In this type of blockchain (Health Insurance) initially tested three types of nodes in a framework that was created earlier. The blockchain framework is Ethereum solidity v4.0.31 but contracts were written. Experiment setup requires Ubuntu 64 bit, Intel i5 processor and RAM, 15.6 GB with 15 Mbps of local area network speeds. After successful completion of experiments, Ethereum gives results like different confirmation time for the different networks as well as smart contracts and is different from the

Ethereum framework. While implementing this proposed method the same framework is used with the help of IPFS for different applications in the same framework environment (Sravan et al., 2018).

Even though blockchain offers a lot of e-services to the government sector again it lacks in offering standards in terms of making a system more reliably secure as well as providing authentication to the users and preserving the data for a long time. In India deploying blockchain in public services is a big challenge. Some of the states in India have started to adopt blockchain-based IT in their state services. This creation of blockchain-based health care and making it a public service may be challenging, but when compared to the above process, the health information or health records of the patient's are brought into important changes in the health public sector concern; records or information has been criticized on account of the centralization process. The problem in the public sector to develop blockchain in all the places is due to the lack of platform availability; this is the major reason why we were not able to develop across the country. However, when we try to develop means the cost will be high. A security system in the blockchain is another criterion while developing in the public sector. This security system requires the following type of security data security (records, information): Physical security, user-provider key-secret key security and risk management. The above types of security ensure trustworthiness of records or information which is stored in the health care sector (Navadkar et al., 2018).

Introducing blockchain technology innovatively with cloud computing as well as IoT technology provides more network capabilities, and computing power may be high.

5.8 Blockchain in Finance

A new force of digital technology is dynamic business models and is progressively turning into an important issue around the world. Blockchain technology is generating significant interest across a wide range of industries in India. The use cases and blockchain fit assessment have also been performed for some banking transactions. Every bank and financial office needs to perform the KYC method one by one and transfer the valid data and documents to the central register. By using a unique ID, banks can access the stored data to perform due diligence whenever customers request a new service within the same banking relationship or from another bank. Having explained about the need for blockchaining in banking and finance, what are all the solutions offered by the blockchaining, uses cases or processes where the blockchaining can play a key role? The main benefits of blockchaining are that it's near real-time, i.e. the blockchain technology enables near real-time settlements of recorded transactions, removing friction and reducing risk. The next benefit is that there is no intermediary, i.e. blockchaining technology is based on cryptographic proof instead of trust, allowing any two parties to transact directly with each other without the need for a trusted third party. The next benefit of blockchaining is irreversibility and immutability, i.e. the blockchain contains a certain and verifiable record of every single transaction ever made. This prevents past blocks from being altered and in turn stops double spending, fraud, abuse and manipulations of transactions (Jani, 2017).

The article examines some of the disruptive changes that are likely to occur in financial services due to rapid technological advances. The article includes a brief mention of regulatory challenges to the adoption of this new technology. Many policymakers are seeking to gain a better understanding of the likelihood that the use of Bitcoins or other cryptocurrencies will gather momentum in their jurisdiction. He previously mentioned that laws and rules may well be programmed into the blockchain itself so that they are enforced automatically. In other situations, the ledger can act as legal evidence for accessing (or storing) data, since it can't be changed (Trautman, 2016).

5.9 Blockchain in Identity Management

The work is based on the distributed ledger technology (DLT), and it leads to novel approaches for identity management to facilitate usage of digital identities. These approaches promote decentralization, transparency and access control in various transactions. Here, they have introduced the fast-growing DLT with IDM and evaluated three proposals,

- Support – this decentralized identity facilitates decentralized identity for all involved entities.
- ShoCard – it helps in face identity verification and online interactions.
- Sovrin – it manages identity in a decentralized manner with the help of permissive DLT.

By using the seminal framework it characterizes the IDM schemes.

There are numerous advantages of the application of DLT with IDM which was proposed earlier:

- Decentralized – the provided identity data should not be owned or controlled by a single authority.
- Tamper-resistant – here the historical activities are transparent.
- Inclusiveness – the bootstrap identity can be conceived, and it reduces the exclusion.
- Cost-effective – by sharing identity data significant reduction in cost is possible.
- User control – the digital control identifier once obtained by the users can be retained without the chance of losing it.

The laws provided for DLT-based IDM schemes were evaluated:

- User control and consent – information should be revealed with the users' consent.
- Disclose minimum for use – provide the data which are essential
- Justifiable parties – the collected information is shared between the parties, and they get the rights to use the information in a transaction.
- Directed identity – for sharing the information the support must be there in public.
- Design for duplication of operators and techniques – every identity scheme must have a solution.

Integration of human – the experience of the user must match the level of users' needs and expectations; then only they can easily interact in the system.

- Regular experience in context – users should have consistency with the experience across the security.

The ambiguity present in the user's elements needs to be addressed (Ali et al., 2016).

The ID system is implemented with the help of the Hyperledger Fabric framework to solve the privacy problem and to make the ID sharing process easy. The current system holds problems such as proxies but blockchains may be a solution for these ID problems.

Hyperledger is an open-source blockchain. It is a platform for confidentiality, resiliency, flexibility and scalability. This system is private; the employees of the Hyperledger Fabric network register in a membership service provider (MSP). It has a ledger subsystem consisting of two components:

- World state – the ledger state is described in a given period of time.
- Transaction log – the current world state is obtained by recording every transaction.

The security concern with this identity management system is sensitive. The personal data stored can be owned by third parties. The proposed system explains the concept of personal cloud where the users reveal the information to the different service provider rather than handling an entire bundle of personal information. It allows the components such as:

- Consensus
- Member services

The client application will be built using bootstrap HTML, CSS and JS deployed on a server.

Instead of the in-memory database, it can migrate relation databases like MySQL or Oracle; instead of limiting this application it can have a single digital ID for all access which will make it still more optimized. It provides a more secure, immutable and user-convenient system (Zheng et al., 2017).

The system propounds a design for BT in identity management and authorization service disruption. The Internet does not have an identifying protocol for the identification of people and institutions. We have all heard about Bitcoin, Ether and other cryptocurrencies, which enable people to anonymously perform secure and trustworthy payments and transactions. In the heart of those cryptocurrencies, there is a blockchain, a decentralized database which records all transactions since their beginning. The entire network, as opposed to a central entity such as a bank or government, is continuously verifying the integrity of it. This way, users do not require a trusted central entity, but security is guaranteed by the strength and computing power of the entire network participating in the blockchain.

Identity management (IDM) refers to board administrative area and standards that create and maintain the user account. Sound identity management and governance are needed to manage identities for online services. Identity management is needed to alter the user provisioning method, enabling new users to get access to online services and de-provisioning users to ensure that only the rightful users have access to services and data. Identity management is classified into different types, and they are independent IDM, federated IDM and self-sovereign IDM. In independent IDM, users do not know their identity record and it can be revoked or misused by the identity provider. In federated IDM, users' accounts are managed independently by an identity provider and no enterprise directory integration is required. In self-sovereign IDM, users should be able to control their own identity. The discovery of this new mechanism creates a secure platform for service providers to authenticate users who do not have a single entry of failure and prevents attacks and leakages of user data. Blockchaining identity management and authentication solution by design are decentralized and distributed, which decreases the deployment and maintenance cost. On the other hand, instead of premise deployment of the blockchain network, blockchain as-as service (BaaS) permits consumers to go ahead with cloud-based solutions to construct, host and make use of their own applications and smart contracts with blockchain.

Blockchain can create a secure platform for online service providers to authenticate users. Besides, this technology could also help to instill trust back in users. Users should have full control over who has the right to use their data and what they can do with it once they gain access (Lim et al., 2018).

5.9.1 Blockchain in Digital Payment

This section provides information on the latest digital payments. It focuses on the operative mechanism of Bitcoin, blockchain technology and describes the scope of this application. In electronic payment systems:

- The mechanism of payments is simplified.
- The procedure for repayment of debt is simplified.
- The difficulties with the conversion from national currencies at bank rates disappear.
- The problem related to money transportation disappears.
- The safety of money is ensured.

This system is: Organizations issue digital currency, create and implement new methods for their distribution and provide conditions for financial transactions. Different electronic payment systems issue their own type of currency. Electronic money refers to the system of storing and transferring both traditional currencies and non-state private currencies.

Classifications of electronic money:

- Smart cards
- Network

Bitcoins are a type of digital currency. Bitcoins are transactions stored in encrypted form with certain conditions for financial transactions. No separate records are maintained regarding the number of Bitcoins. In cryptocurrencies, public and private keys are used to transfer currency from one person to another. It reaches a conclusion with the sole proprietor of the key able to control the Bitcoins (Tschorsch & Scheuermann, 2016).

FinCEN Bureau from the USA proposed various measures to monitor transactions across different cryptocurrency exchanges. It also recommended measures to curb the exchange rate differences and the revenue generated by miners. In order to collect income tax, the income from Bitcoins is treated as property. The Australian Taxation Office (ATO) encouraged cryptocurrencies since their law does not restrict their citizens in choice of currency. Airports and several shops accept them as legal tender. This allows anonymity in tickets, avoiding intermediaries. Since the population is lower, this method is very comfortable. From the end of 2013 the Swiss government has considered cryptocurrencies as another foreign currency. The incorporation of blockchain technology was widely adopted in their financial services like trade exchanges, banks and exchanges. Japan is identified as the origin of the protocol of cryptocurrencies. Here Bitcoins are approved as legal tender. It is also the first government to impose regulatory controls for these transactions. Recently the Chinese government banned the usage of Bitcoins due to the issues with taxation and communist ideologies. However, blockchain technology is massively adopted in various government departments which are helping them to avoid intermediaries (Anastasia, 2018).

Payment with card transactions concludes after much verification of information interchanges between the merchant, cardholder, issuing bank, a merchant bank and any intermediate card processors. Blockchaining technology is used to record all the transactions. In blockchaining technologies, we have two types of blockchaining, and they are private blockchaining and public blockchaining. Private blockchain makes use of a linked list with inbuilt hash pointers that are used to record secured transactions in a well-defined manner (Godfrey-Welch et al., 2018).

In a public blockchain, any user can join, consolidate and publish transactions. When all nodes are not known to each other then it is a permissionless or stateless blockchain. A recognizable implementation of a public blockchain is Bitcoin. Public blockchains are considered to distribute and maintain more massive ledgers, thereby requiring more computation resources. A private blockchain provides network needs and should validate by either the network operator or by a collection of rules place in situ by the network operator. When all writing nodes are known then it is a permissioned blockchain (Jayachandran, 2017).

The main benefits of blockchaining in digital payments are reduced transaction participants, reduced transaction processing time, utilization of a single encrypted transaction ledger, increased data integrity and reduced transaction fees. The major constraints and risks are imposed delay, performance (runtime versus real time), throughput, adoption, limited block size and card storage. Coming to the risks: the main risks are lost or stolen credentials, network availability, network integrity and trust.

In summary, by applying this decentralized technology there is a possibility to reduce the burden of almost all the government activities significantly. This may lead to total virtual governance in which the entire administrative machinery can discharge its duties without any manual intervention that can be audited at any time for any reason. The permissive blockchain and stateless blockchain, when combined suitably, may lead to a fully democratic government in which there is no central power of authority, totally eliminating corruption, bias and over-exercising of power. Also, the cost of running government activity would be slashed drastically that may in turn benefit the people of the country by heavily reduced direct and indirect taxes. This will lead to most of the money earned by the government for the reduction of unemployment and other social welfare schemes. In another 20 years, many fast-growing countries may join the developed nation bandwagon due to this disruptive technology.

References

Adida, B. (2008, July). Helios: Web-based open-audit voting. In *USENIX Security Symposium* (Vol. 17, pp. 335–348).

Ainsworth, R. T., & Shact, A. (2016). Blockchain (distributed ledger technology) solves VAT fraud. Boston Univ. School of Law, Law and Economics Research Paper, (16–41).

Ainsworth, R. T., & Viitasaari, V. (2017). Payroll tax & the blockchain.

Ali, M., Nelson, J., Shea, R., & Freedman, M. J. (2016). Blockstack: A global naming and storage system secured by blockchains. In *2016 {USENIX} Annual Technical Conference ({USENIX}{ATC} 16)* (pp. 181–194).

Anand, A., McKibbin, M., & Pichel, F. (2016). Colored coins: Bitcoin, blockchain, and land administration. In *Annual World Bank Conference on Land and Poverty*.

Ananth, C., Karthikeyan, M., & Mohananthini, N. (2018). A secured healthcare system using private blockchain technology. *Journal of Engineering Technology*, *6*(2), 42–54.

Anastasia. (2018, May 30). Top 5 countries embracing the blockchain technology. Retrieved from Yogita Khatri. (2019, Janurary 17). Wyoming blockchain bill proposes issuance of tokenized stock certificates. Retrieved from https://www.coindesk.com/wyoming-blockchain-bill-proposes-issuance-of-tokenized-stock-certificates.

Ayed, A. B. (2017). A conceptual secure blockchain-based electronic voting system. *International Journal of Network Security & Its Applications*, *9*(3), 01–09.

Beck, R., Avital, M., Rossi, M., & Thatcher, J. B. (2017). Blockchain technology in business and information systems research.

Budish, E. (2018). The economic limits of Bitcoin and the blockchain (No. w24717). National Bureau of Economic Research.

Chaum, D. L. (1981). Untraceable electronic mail, return addresses, and digital pseudonyms. *Communications of the ACM*, *24*(2), 84–90.

Daniel, J., Sargolzaei, A., Abdelghani, M., Sargolzaei, S., & Amaba, B. (2017). Blockchain technology, cognitive computing, and healthcare innovations. *Journal of Advances in Information Technology*, *8*(3).

Evans, D., & Paul, N. (2004). Election security: Perception and reality. *IEEE Security & Privacy*, *2*(1), 24–31.

Godfrey-Welch, D., Lagrois, R., Law, J., & Anderwald, R. S. (2018). Blockchain in payment card systems. *SMU Data Science Review*, *1*(1), 3.

Hanifatunnisa, R., & Rahardjo, B. (2017, October). Blockchain based e-voting recording system design. In *2017 11th International Conference on Telecommunication Systems Services and Applications (TSSA)* (pp. 1–6). IEEE.

Heston, T. (2017). A case study in blockchain healthcare innovation.

Jani, S. (2017). Scope for Bitcoins in India.

Jayachandran, P. (2017, May 31). The difference between public and private blockchain. *IBM Blockchain Blog*.

Lai, C. C., & Liu, C. M. (2019). A mobility-aware approach for distributed data update on unstructured mobile P2P networks. *Journal of Parallel and Distributed Computing*, *123*, 168–179.

Lim, S. Y., Fotsing, P. T., Almasri, A., Musa, O., Kiah, M. L. M., Ang, T. F., & Ismail, R. (2018). Blockchain technology the identity management and authentication service disruptor: A survey. *International Journal on Advanced Science, Engineering and Information Technology*, *8*(4–2), 1735–1745.

McCorry, P., Shahandashti, S. F., & Hao, F. (2017, April). A smart contract for boardroom voting with maximum voter privacy. In *International Conference on Financial Cryptography and Data Security* (pp. 357–375). Springer, Cham.

McMurren, J., Young, A., & Verhulst, S. (2018). Addressing transaction costs through blockchain and identity in Swedish land transfers.

Mendes, D., Galvão, H., Eiras, M., & Lopes, M. (2017). Clinical process in blockchain for patient security in home care. *Journal of the Institute of Engineering*, *13*(1), 37–47.

Navadkar, Vipul H., Nighot, Ajinkya, Wantmure, Rahul. (2018). Overview of blockchain technology in public /government sectors. *International Research Journal of Engineering and Technology, 5*(6).

Nguyen, B. (2017). Exploring applications of blockchain in securing electronic medical records. *Journal of Health Care Law & Policy, 20*, 99.

Noizat, P. (2015). Blockchain electronic vote. In *Handbook of Digital Currency* (pp. 453–461). Academic Press.

Pokrovskaia, N. N. (2017, May). Tax, financial and social regulatory mechanisms within the knowledge-driven economy. Blockchain algorithms and fog computing for the efficient regulation. In *2017 XX IEEE International Conference on Soft Computing and Measurements (SCM)* (pp. 709–712). IEEE.

Ramya, U. M., Sindhuja, P., Atsaya, R. A., Dharani, B. B., & Golla, S. M. V. (2018, July). Reducing forgery in land registry system using blockchain technology. In *International Conference on Advanced Informatics for Computing Research* (pp. 725–734). Springer, Singapore.

Sadiku, M. N., Eze, K. G., & Musa, S. M. (2018). Blockchain technology in healthcare. *International Journal of Advances in Scientific Research and Engineering, 4*.

Schwanke, A. (2017). Bridging the digital gap: How tax fits into cryptocurrencies and blockchain development. *International Tax Review*.

Sharma, P. K., & Park, J. H. (2018). Blockchain based hybrid network architecture for the smart city. *Future Generation Computer Systems, 86*, 650–655.

Shilpa, Sharma, R., Singh, S. (2018). Big data analytic on blockchain across healthcare sector. *International Journal of Engineering and Technology (UAE), 7*(2.30), pp. 10–14.

Singh, M., & Kim, S. (2018). Branch based blockchain technology in intelligent vehicle. *Computer Networks, 145*, 219–231.

Sravan, Nukala Poorna Viswanadha, Baruah, Pallav Kumar, Mudigonda, Sathya Sai, and Phani, Krihsna. K. (2018). Use of blockchain technology in integrating health insurance company and hospital. *International Journal of Advances in Scientific Research and Engineering, 9*.

Swan, M. (2017). Anticipating the economic benefits of blockchain. *Technology Innovation Management Review, 7*(10), 6–13.

Themistocleous, M. (2018). Blockchain technology and land registry. *The Cyprus Review, 30*(2), 199–206.

Thomas, R., & Huang, C. (2017). Blockchain, the Borg collective and digitalisation of land registries. *The Conveyancer and Property Lawyer* (2017), 81.

Trautman, L. J. (2016). Is disruptive blockchain technology the future of financial services.

Trueb, B. A. (2013). Estonian Electronic ID–card application specification prerequisites to the smart card differentiation to previous versions of EstEID card application.

Tschorsch, F., & Scheuermann, B. (2016). Bitcoin and beyond: A technical survey on decentralized digital currencies. *IEEE Communications Surveys & Tutorials, 18*(3), 2084–2123.

United Nations. Economic Commission for Europe. (1996). Land administration guidelines: With special reference to countries in transition. United Nations Pubns.

Vos, J. (2016). Blockchain-based land registry: Panacea illusion or something in between? In *IPRA/CINDER Congress*, Dubai.

Vos, J., Beentjes, B., & Lemmen, C. (2017, March). Blockchain based land administration feasible, illusory or a panacea. In *Netherlands Cadastre, Land Registry and Mapping Agency. Paper Prepared for Presentation at the 2017 World Bank Conference on Land and Povertry*, The World Bank, Washington, DC.

Wang, B., Sun, J., He, Y., Pang, D., & Lu, N. (2018). Large-scale election based on blockchain. *Procedia Computer Science, 129*, 234–237.

Wanitcharakkhakul, L., & Rotchanakitumnuai, S. (2017). Blockchain technology acceptance in electronic medical record system.

Wijaya, D. A., Liu, J. K., Suwarsono, D. A., & Zhang, P. (2017, October). A new blockchain-based value-added tax system. In *International Conference on Provable Security* (pp. 471–486). Springer, Cham.

Wright, A., & De Filippi, P. (2015). Decentralized blockchain technology and the rise of lex cryptographia. Available at SSRN 2580664.

Zheng, Z., Xie, S., Dai, H., Chen, X., & Wang, H. (2017, June). An overview of blockchain technology: Architecture, consensus, and future trends. In *2017 IEEE International Congress on Big Data (BigData Congress)* (pp. 557–564). IEEE.

Zheng, Z., Xie, S., Dai, H. N., & Wang, H. (2016). Blockchain challenges and opportunities: A survey. Work Pap.–2016.

Chapter 6

Blockchain and Social Media

Saugata Dutta and Kavita Saini

Contents

6.1 Introduction

A blockchain can be defined as an increasing list of blocks where each block is linked with the hash of the previous one. A block consists of a collection of transactions, time stamps and the hash of the previous block. Blockchain is a distributed ledger where transactions are stored in blocks and spread across a peer-to-peer network where every node holds a copy of the ledger. The value of cryptography hashing and a distributed ledger makes it more unique than other technologies in that it is highly secured and almost impossible to compromise. The transparency in information for every node or participant is a property of blockchain which shows information cannot be modified which reduces fraudulent data and tampering with data which helps in building trust. Blockchain is customizable where actions can be triggered once the contextual conditions are met. Blockchain works on a peer-to-peer network, and there is no third party or central authority involvement.

Encryption is a base concept for blockchain. It can be seen as immutability of data which is also distributed. The history starts during the late 1970s when the concept of a Merkle tree was introduced which is a binary hash tree. The conceptual tree consists of a leaf node with a hash and every non-leaf node has a cryptographic hash of its child nodes. This helped in the secure verification of large data. This conceptual model was invented by Ralph Merkle in 1979. In the early 1990s a blockchain-like technology was introduced by Stuart Haber and W. Scott Stornetta, working on secure blockchain. In 1992, documents could be stored in a single block with the help of the Merkle tree design. The actual inception of blockchain was conceptualized by Satoshi Nakamoto in 2008 when he introduced the peer-to-peer cash system. This triggered the launch of a currency in 2009 which is a digital cryptocurrency with an underlying layer of blockchain technology. The digital cryptocurrency known as "Bitcoin" came to existence. Satoshi Nakamoto mined the first Bitcoin. Hal Finney who was the programmer for Bitcoin was the first to receive 10 Bitcoins from Satoshi Nakamoto.

The year 2014 experienced a probability to use the blockchain technology as a legitimate use of payments. This is the starting period where investors started focusing on this technology. This is the time when smart contracts and decentralized apps were conceptualized. This stepled to the realization of the benefit of running decentralized apps and smart contracts. Ethereum is a public blockchain developed in open source and decentralized whereas the cryptocurrency is named "Ether". Ethereum with a smart contracting layer gave the advantage of including various parameters and also validating various levels over the transactions. This contracting layer also helped various parameters and improved upon the overall interactions on the transactions and contracts. Blockchain technology is transparent, secured and there is no third party or central authority intervention. It has the ability to make an organization that uses this technology more transparent, democratic, efficient, decentralized and secure. Within the next five to ten years blockchain will disrupt various industries. Some of the industries where blockchain has already started include banking services intended for those who have no access to the banking system that we use presently. Similarly Bitcoin allows users to send money. Barclays has already started using blockchain to some extent. Blockchain is used in supply chain management where users store data in decentralized locations and records are kept in a secured manner which provides various benefits like minimizing cost and labor. Some of the startup blockchain companies like Provenance, Fluent, Skuchain and Blockverify are working to improve supply chain management. In forecasting industries, blockchain is expected to change the traditional methods of research, consulting, forecasting and analysis. In networking and IoT, IBM and

Samsung are using a new idea called "adapt" which will use blockchain technology and create a distributed network of IoT devices. In the insurance industry, the core is trust management. With the help of blockchain, technology ensured person identity. Blockchain can be used to verify many types of data in insurance contracts like the insured person's identity. Blockchain smart contracts can be integrated with oracles bearing real data. A blockchain project called Aeternity is developing digital applications for the insurance industry. In private transport and the ride-sharing industry, blockchain technology is being used to create a platform for riders and owners where riders can access the services by meeting a mutual terms and condition without a central authority. Blockchain startups like Lazoos and Arcade City are working in this area. In the online data storage industry, blockchain allows storage to be more secure and robust against attacks. Storj is an example of decentralized cloud storage. In the charity industry, blockchain can be used to address inefficiencies and corruption. It can help to track that donations are going to the right hands. BitGive uses blockchain technology to receive funds using a distributed ledger. In voting, blockchain can be used for a number of services like the verification of identity, registration of voters and also counting the votes. It helps in counting only the legitimate votes, no votes are changed or moved, and creating an immutable, publicly viewable ledger will make a massive step forward to make elections more fair and democratic. Democracy Earth and Follow My Vote are aiming in this field to create a blockchain-based voting system. In the health care industry, blockchain is used to store and share sensitive data with authorized doctors and patients. This will help in data security and improve in accuracy and speed in diagnosis. Gem and Tierion are working in the health care data space. In the energy management industry, TransactiveGrid allows customers to buy and sell energy from each other in a decentralized way without involving the public grid or a trusted private intermediary. In online music, blockchain can be used by fans to pay musicians directly. Smart contracts can be introduced which can solve licensing issues and better catalog songs with their respective creators. Mycelia and Ujo are blockchain-based platforms made specifically for the music industry. In the retail industry, blockchain utilities creates a lobby where sellers and buyers meet and can deal as per the requirements without the intervention of a central authority. The two startups OpenBazaar and OB1 use blockchain in the retail space. In the real estate industry, some challenges can be overcome like lack of transparency, fraud, bureaucracy and mistakes in public records. Blockchain technology can help in reducing the need for documents, building accuracy and verifying ownership and also speeding up transactions. Ubiquity is a blockchain-based secured platform for real estate record keeping which is an alternative to legacy paper-based systems. In the crowdfunding industry, many organizations are benefited especially the startups by raising funds using blockchain smart contracts which eliminates the need for a third party. New projects can release their own tokens that can be later exchanged for services, cash or products.

6.2 How Blockchain Works

Blockchain is a peer-to-peer technology where the integrity of digital information is protected. Blockchain is defined as a decentralized ledger of transactions over a peer-to-peer network. Blockchain offers secured and transparent transactions to all parties involved. The blockchain ledger records every sequence of transactions from beginning to end whether it is many or one. On each transaction, it is put into a block and each block is connected to the one before and after [2]. The lists of transactions are blocked together and the fingerprint of each block is added to the next, thus creating an irreversible chain. Blockchain works with all types of transactions, and it is distributed, permissioned and secured. A block usually consists of the current transaction information and hash of the previous block. A hash function takes any digital media and runs an algorithm on it to produce a fixed-length digital output known as a hash (Figure 6.1).

This fixed-length output is smaller than the original input. When even a single bit of digital media is changed, the hash function also changes as compared to the original one. All blocks are numbered starting from 0. The first block numbered 0 is known as the genesis block.

The distributed ledger nature of blockchain makes it more secure. The data are not stored on a centralized database but across a wide range of computers known as nodes. If any of the blocks is tampered with, it causes the hash of the block to change which will make all the subsequent blocks invalid. Nowadays computers and technology are pretty fast, say for example if any of the blocks is tampered with and all the subsequent block hash are recalculated, then there is a chance the blockchain will be compromised. In order to mitigate this issue, there is a concept called proof of work (POW). It actually slows down the creation of blocks (Figure 6.2).

So if any of the block is tampered with, then proof of work needs to be calculated for all subsequent blocks. Now proof of work (POW) is a mathematical design where a pseudo-random number is presumed and joined with block data to calculate the hash, producing a result that equals a given criteria, for example a hash with a prefix of three zeroes. Once the result with the given condition is found, the result is verified by other nodes and the miners are rewarded with digital currency

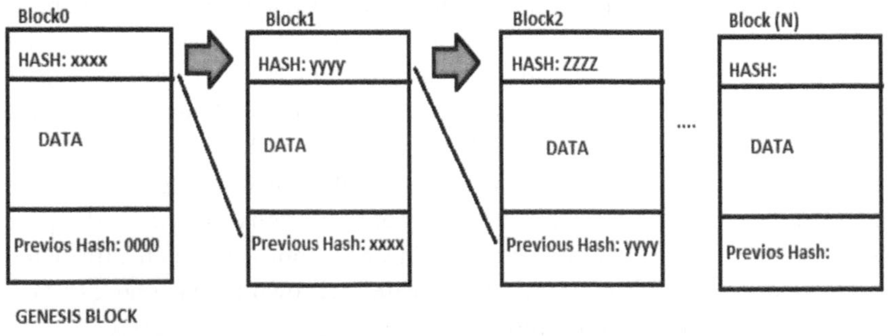

Figure 6.1 Blocks in blockchain.

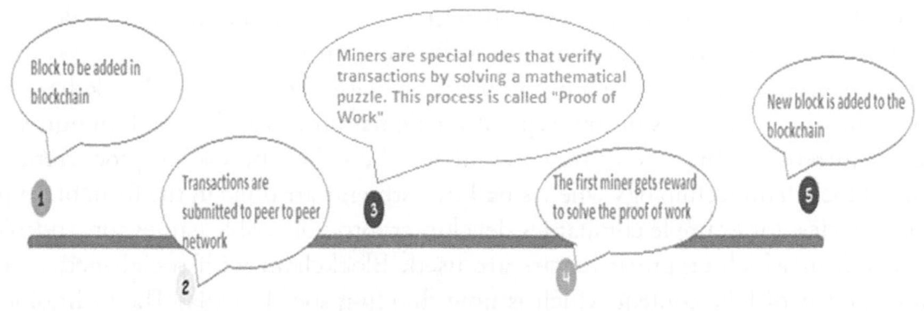

Figure 6.2 Adding a block to the blockchain.

[3]. As mining is a guessing game, miners are special nodes with special hardware or computational power who take part in this guessing game known as proof of work and are rewarded with digital currency on successfully guessing and adding a block in the blockchain. The miners with more computational power may succeed more often, but due to the law of statistical probability, it is highly unlikely that the same miner will do so every time. Using a consensus algorithm, you can actually agree one particular state of blockchain in the entire network. So if a new block is added in a blockchain, who will add it can be defined with the help of a consensus algorithm like proof of work (POW).

6.3 Blockchain in Social Media

To start with, the social media system is a walled garden which is protected and interestingly exploited. In one way we can say that we all are products which ironically social media monetize. We certainly have no option but trust social media to handle our data. There have been incidents where information data were compromised. There are a number of criticisms of public disclosure by social networking sites. Importantly, the information in the social media has no validation or authenticity, it totally depends upon the face value. However with the principles of blockchain, a mechanism exists by which we can deal with social media from a different angle altogether. Here we can participate directly without the intervention of third-party trust and we profit from the success of what we are posting. Now there are certain factors like decentralized social media; next is to reinvent the incentive structure of social media and also the validation and authenticity of posts. Social media is undergoing a metamorphic shift at this moment. Social media will experience consequences from the disruption of blockchain technology [4].

There are certain factors which influence shifting of social media with underlying blockchain technology. Verification of online identities is in increased demand. Fake IDs and vulnerability to fraud can be mitigated where the identities of customers can be verified using blockchain technology and smart contracts. Similar to the case of identities, marketplaces can also be verified using blockchain technology, where vendors are verified which will make marketing simpler and improve the return of investment and will also gain potential company growth. Cryptocurrencies can be a good combination with the advent of blockchain in social media. Posts, likes or any type of participation can yield small amounts of cryptocurrencies. There is also a demand for the collectibles of cryptocurrencies using blockchain technology where small investments are done in the form of cryptocurrencies, for example companies develop crypto-collectible games for Android and IOS in which cryptocurrencies are used. Blockchain with social media can stop the use of fake content which is now flooding social media. The technology will help to post traceable data and verified contents. This will help marketing teams and brand managers. Journalistic social media faces challenges on the media content that audiences rely on such as source reliability, excessive liberty in privacy, fake news and failure to regulate the structure and flow of information [5]. Although social media has given us lots of useful information, many of this is untraceable, has integrity issues and is unreliable and unethical.

6.4 Blockchain-Based Social Media (A Revolution)

A social media revolution has already started with various startups. The timespan required to introduce social media to the world is more or less the same as the timespan expected to expand the blockchain-enabled social media. This is in fact being used as the advantage of earning money, censorship, immutability, transparency, creditability, authenticity and reliable platform. The application is decentralized with no central authority control and transparent. The advantage of getting situated in this platform will be highly creditable, as with other social networking sites where the first subscriber of Twitter has a million followers and so with YouTube, where the first video yields millions of likes; the sooner the better. In this authentic platform, people will find each other in blockchain-enabled social media and can later follow each other on Twitter or other social networking links rather than vice versa. This will help in sales and marketing. Social media is actually not social and is basically a push mechanism. Data leaks, unreliable content and news, public disclosure and advertisers appealing for attention are some of the reasons why people started realizing the value of data and the risk of posting in public platforms. This platform is on the rise and also ready to change the social media infrastructure and bring a different paradigm altogether. Some of these following issues can be addressed:

- **Fake news**: The advantage of using authentic content and source data verification will help to fight this problem. The content creators are also verified. The approach is versatile towards profitability on personal data, control and privacy.
- **Advertisement**: Where the user has the control to see ads at their own wish. This will help in controlling advertisement frauds where there is wastage of money. The users may be rewarded if they wish to see the advertisement.

- **Rewards**: With the help of smart contracts, creators are rewarded for views and likes for quality posts. In comparison in other networking sites like YouTube the compensation is too little and is only triggered with huge view numbers.
- **Privacy**: Users can be anonymous if they want or may log in to their secured account as per choice. The anonymity will help them not to be censored, traced or tracked where in some countries social networking is blocked. The benefits of social media with this technology will eliminate blocking.

The strongest argument made for this technology is that you have the control over your data. This technology will be helpful not only for the brands but for individuals too because the users select the brands and they get paid for it. An individual's digital identity is worth everything. Imagine a celebrity's digital identity on a social networking site is valuable and a money earner. The answer remains to us for our likes and views which earn them money. That is one of the major reasons why the decentralization movement is powerful and we all will be a part of it. The tokenized business model will help to generate token rewards for photo uploads, views and likes. This will start a community business.. That is where the future is going, and this is what the technology enables. Although the ideas and concepts were there earlier, the technology execution was not possible, but now with the advent of various latest technology models it is possible with various startup companies focusing on these models.

6.5 Social Media Opportunities in Blockchain

Social media opportunities are very broad and apart from these it's a huge platform. Most of us or in fact all of us have a social media account. There have been several surveys and research conducted where an average teenager spent eight to nine hours per day on interactions. The issue here is the privacy, where marketing companies conduct promotions and advertisements for audiences and hence newsfeeds are loaded with advertisements. The opportunity to earn money for end users for posts is not possible unless views are exceeded. The present scenario of social media is centralized where users' privacy and data security are at stake. Blockchain-based social media is the only answer to this issue where data privacy and security are taken care of and there is no control by a third party. Blockchain-based social media also use digital currencies to reward creators and viewers. They are a great alternative to current traditional social media platforms [6].

Here are some current platforms of blockchain-based social media.

- **Steemit**: This is the most common blockchain-based social media and needless to say it is decentralized; users can experience the same features as on Facebook. Users will have better data privacy and security. The creators are rewarded with digital currencies and for all in-platform transactions. Steemit works with upvotes and downvotes. The earning mechanism works based on the number of upvotes it receives for the content. The more the user is engaged, the more he or she earns. It is the most successful blockchain-based social media. The distribution of rewards is 50% to authors and the rest, 50%, to voters. The 50% of reward given to authors is further divided as 50% of the reward is given in Steem powers and rest, 50%, in Steem dollars. Steem power rewards are distributed over two years of weekly payments. This is done to bring stability to the market, not just to speculate and make money but to help in continuous engagement. Steem powers are given to winners to curate content and earn Steem rewards. The sign-up process takes some time, as this ensures that the participants are real people. In this platform individuals are rewarded for creating and curating content. Steem provides opportunities for creators, curators, remitters, merchants, shoppers, market makers, commenters, entrepreneurs, bloggers, referrers, community leaders and Internet readers.

- **Obsidian**: It provides a collection of blockchain-based services and apps. But the main purpose of obsidian is a highly secured blockchain-based messenger. Due to its peer-to-peer nature, all messages are encrypted end-to-end and you have full control over the data, files and photos you share. There are features like timely auto deletion, and there is no intervention of a third party. An individual can be assured that there is no breach in data privacy and security. On a decentralized network of blockchain the encrypted private messenger operates.
- **Earn**: It's an application quite similar to LinkedIn. There are some differences to it. Earn enables the user to receive messages which are paid, and the user can also link with people of the same skills. One can make money through the paid email system, inspiring users to pay for responses. One can create an auto-reply to mail as well.
- **Indorse**: This platform is built on Ethereum blockchain. This is also similar to LinkedIn and sets profit from the skills when an account is created. The contribution is rewarded and users have full rights to their content. In Indorse, users can benefit from some technical skills and end up getting business or job when reviewed by experts. This also allows smart contracts and rewards users with digital currencies. All connections are validated, authentic and reliable. This is a reliable channel to endorse skills and also display them. Each user either can be a claimant or a moderator. A user once creates an account and can claim the profile they upload. The same is reviewed by others, called proof of stake. Once the claim is accepted and authentic, the user's score is elevated. This helps in the KYC process, freelance services, hiring, advertising and market predictions. The digital currencies are used by the advertisement firms to buy space in the social media. Score token are used for posting, views and updates to the member profile.

- **Social X**: A similar app to Instagram and Facebook. On a highly secure platform power-driven by blockchain, one can post videos, photos and updates. This has a feature of license management where you have full control of your photos and helps to sell them to other users as well. Since it's a decentralized network with blockchain in play, the user has full security and control over public disclosure. SOCX tokens are used to exchange photos among the users. It has a highly secured messaging feature where one can exchange messages and digital currencies. One can earn rewards by sharing content. Holistically, it is a blockchain-based social media platform which has all features. Social X is a distributed media platform running on blockchain and interestingly powered by individuals. The Social X infrastructure is arranged into three layers. One is for social data, the second is for transactional data and the third is for distributed media. The Social X eco system allows the distribution of advertisement revenue back to the community in a unique reward system.
- **Enlte**: A decentralized platform in social media which claims to solve real-life problem without an intervening central authority or governance. The social networking platform is location-based with more secured and better networking. It is said to be the location-based small world network. This will help to find accurate and correct information. One can share experiences and can have a qualitative value with authentic data. This platform is advertisement-free. One can also be anonymous and operate on the platform. All experiences are geo-tagged. If one posts an experience, the same will be stamped with geo location and broadcasted to other users in the same geo location. Users will be rewarded for the awareness and creating a supportive environment. Each user has the right to vote and approve the block and hence this can be mined as it doesn't require any computational power. This will help in generating revenue like mining and earning coin. Users have the prerogative to buy and sell Enlte coins from the inbuilt exchange.
- **Voice**: It's a blockchain-based social media platform which ensures the authenticity of the userbase. It's a platform where good content is circulated in the community. The contents are shared and promoted and users get direct benefits for their creativity and ideas. All engagements in Voice are transparent and public, and there are no hidden agendas or algorithms. Information security and privacy is a core element for Voice. When one's post goes live and popular, Voice tokens can be earned. It is built on EOS public blockchain.
- **Sapien**: It is a social networking platform using blockchain technology built on public Ethereum. It's also similar to Facebook, which is a social news platform. Sapien uses SPN tokens which serve as the backbone of tokenized economy. Quality content is incentivized by rewarding users with SPN tokens. These tokens can be used to buy physical and virtual goods. A user's content is validated by users of the same domain accumulating reputation points that reflect the expertise. This Sapien network will allow users to limit the spread of fake news.

- **Sola**: It is an amalgamation of social network and media. It was designed in order to address the unfair monopoly of traditional social media, to give users an equal voice [7]. Users' posts are either endorsed or skipped. When endorsed, the content will be displayed to mass users; hence there are no concepts of share and like systems for promotion. It uses a viral geo system, where users' content gets shared automatically to the closest users measuring the proximity. SOL is a utility token that is used in app transactions, rewards users, advertisements and purchase services.
- **Ong.Social**: Is a blockchain social dashboard and also provides social rewards with cryptocurrency. It actually runs with Ethereum and Wavesplatform. It helps users to get a chance to earn money on every post. It's an incentivized social dashboard to connect and control social media accounts and also uses a decentralized social network where information will not be censored and restricted. One can post content and share it on other social media accounts and so on. Here contents are never removed or banned. It uses a gravity algorithm, which is used to identify and measure viral contents to track a number of people in a short amount of time. The measurements are made through positive and negative gravity. Positive gravities are rewarded while negative gravities are not rewarded but at the same time neither is punished. It uses dual blockchain concepts both on waves and Ethereum. It uses onG Coins for the ecosystem.
- **Minds**: Is a blockchain based social networking service where users are rewarded for their quality content and contributions. It is a peer-to-peer and distributed network. This is built in Ethereum blockchain. In this platform one can advertise a contribution with the use of tokens. In this platform, data privacy and security is taken care of. Users own their data, and there is no security breach. One token can be equivalent to 1,000 views. It also has a feature of premium subscription where the user needs to pay five tokens a month to get exclusive content and verification [8].
- **SoMee**: Redefines the traditional blockchain media by empowering choice, inspiring community, respecting privacy and rewards to contributors. It is a blockchain-based social media platform which uses ONG cryptocurrency. It helps users to create communities without security breach, have total control of data and earn rewards of social media channels [9]. It has a wallet system powered by cryptography private keys where digital payments are made. The payment system has full transaction records which cannot be tampered or compromised. SoMee respects private information and gives each user the option to share in revenue generation. Users can earn block rewards and can receive the USD value of the rewards generated from users' contributions. SoMee puts your data in your control completely, and one can never lose any reward or channel and apart from this there will be no censorship or shutdown. This also benefits users to create, view, connect and share information without the fear of being abandoned, and users have full control of

their data. Users can see all social feeds and also have the facility to share without disturbing the existing system. It's an incentivized social dashboard to connect and control social media accounts and also use a decentralized social network where information will not be censored and restricted. Here contents are never removed or banned. It uses a gravity algorithm, which is used to identify and measure viral content to track a number of people in a short amount of time. The measurements are made through positive and negative gravity. Positive gravities are rewarded while negative gravities are not rewarded but at the same time not punished. It uses dual blockchain concepts both on waves and Ethereum.

- **Smoke**: It is a decentralized cannabis network. This is a blockchain built for the cannabis community [10]. This is perhaps the first blockchain for cannabis services. This is resistant to censors, and the database is immutable. Users can earn rewards and have business access without the disadvantage of false identity and content. It uses a social consensus algorithm to mine for reward transactions. The contents can be reviews, cannabis strains, articles, etc. Voting rights and high-quality content rewards are done through special utility coins called SMOKE cryptocurrency.
- **Alfa**: It is a blockchain-based social media platform which has enhanced privacy and a user-friendly approach. It fosters digital sharing and gives the opportunity to become a small entrepreneur without any central authority. It has premium advertisement for various promotional activities and are all time-based. Users will have the prerogative to see the ads as per their choice and are rewarded for time spent watching ads. All the content postings are time-based and in chronological order. The users can schedule posts on a time frame; the posts can also be self-destructible. Breaking the walled garden as against the traditional social media system, users are encouraged and given value for the post while giving full control to users. It creates an eco-system where users have the chance to be micro-entrepreneurs. This will provide greater possibilities for everyone. Buyers and sellers and benefited in this platform by services such as ride sharing, home sharing, information sharing and other services where transactions are kept private. The content navigation system is very appealing and can be accommodated on any screen. It also has an alpha watch where all these apps are clubbed together like social network, digital wallet and smart phone to have a different unified experience altogether.
- **APPICS**: It's a blockchain-based social media application covering various categories like sports, arts, culture, lifestyles, fashion and so on. Users are rewarded for content posting. The reward token is cryptocurrency-based and is paid for creative content and at the same time helps users to earn a share from the overall revenue. This platform helps to monetize the content created and respect users who devoted time and energy to create content to keep the site alive. The users who like the contents are also rewarded, creating a

win-win situation for both creators and viewers. You can send direct transactions to another user's account which are totally secured and take a matter of seconds with no fees and no delays. There are 16+ categories.

- **Peepeth**: It is a decentralized social media like Twitter. Peepeth runs on Ethereum blockchain. It runs like any other normal website except the smart contract and data stored are open source. Peepeth reads and writes from the smart contracts. One can use bulk posting, to post all data like posts and follow in a single cheap transaction [11]. There is an option for free peeping where a user can post for free if they have peeped several times and have good social standing. It has features like twitter cross posting, bulk posting and mail notifications. It has a convincing social media verification process, where the user needs to post the Ethereum address to an external social media account which the user claims. The link will then be verified by smart contracts and, once done, the external social account will be linked with the Ethereum address. It gives the user control over their own legacy. Data are saved on a public blockchain where data are immutable. It has great features like resistance to SPAM, transparency, verifications, authentic contents and monetization.
- **Mastodon**: It's a blockchain-based social network and open source-based application which is similar to Twitter or Tumblr where users post photos, videos and messages and follow, share or like and data are saved on a blockchain platform. Messages use a limitation of 500 characters and are ordered chronologically. It is decentralized, and there is no intervention from any central or third-party authority. When a user creates their own version of Mastodon it is known as instance. Users can create Mastodon instances with their own set of rules and have full ownership over it [12]. Users can follow each other on cross instances and can communicate with other users seamlessly from other instances. Private instances have the freedom to choose to communicate with other users. It offers anti-abuse tools to moderate over the instances. This type of social media cannot be sold, blocked or bankrupted. Users have the ability to join any community they want and communicate. Mastodon with its huge registered user base is growing fast among blockchain social media platforms.
- **Steepshot**: It's a social media application based on blockchain technology where the user can earn rewards when uploading quality content. This is built on Steem blockchain. It is also resistant to censorship. It is perhaps the first blockchain-based application in Steem blockchain for sharing photos and videos primarily. Users can earn Steem dollars. Similar to other blockchain-based social media applications the user can share posts to other social media sites and promote posts while earning rewards. It consists of a mobile wallet where one can conduct transactions to claim rewards and check one's balance without going to Steemit account. It is equipped with push notifications for upvotes, comments and likes and also can achieve updates from a specific

user. With the help of an NSFW tag, a user has the ability to post gore, pornographic, horror and violent contents. It has the capability to search content based on certain topics or subjects. Full-screen mode, photo editing and tagging are some of the enhanced features [13].

- **Dtube**: Is an alternative to YouTube, where users can earn in cryptocurrencies when a video is uploaded. The platform is blockchain-based. The uploaded videos can earn rewards for seven days [14]. The videos can be upvoted or downvoted depending upon the quality of the content. The more upvotes, the more a user can earn. It's built on Steem blockchain. Like YouTube, users can create channels and subscribe to channels. It is resistant to censorship, because of its decentralized nature. It is a fair platform with full transparency, where there is no hidden algorithm. It's advertisement-free, however the user has the freedom to put advertisements into their videos, but take their own risk of losing subscribers from their channel. Dtube uses IPFS for decentralized file storage. The content is hashed, and the hash becomes the identifier of the uploaded file. This can also be rehashed and compared to check the integrity of the file. This is also called DHT, which is known as distributed hash table.

6.6 Future Ahead

Blockchain helps in securing content on social media. This might not be a concern for most, but for those who have concerns, they might not be subscribers to social networking sites at all. Social networking sites have their own terms and conditions, promotion ideas and collaboration. Crowdfunding companies like Indiegogo and Kickstarter help startup organizations to raise funds through digital currencies or ICOs.

Thus, the quality of blockchain-based social networking sites help to raise funds securely and without corruption. Without the external payment mechanism, the network that is running on cryptocurrency can support crowd sales. Blockchain-enabled social media will have an intense impact on widespread platforms like Twitter, Facebook, Instagram and Snapchat. Marketers in social media should explore and adopt these new blockchain-based social networking lobby to be an early bird in this opportunity. In the absence of any centralized authority, users on these networks benefit from greater privacy [15]. In turn, this upholds the freedom of speech and expression, relieving users of the torments of being prosecuted for their thoughts on social media. Most decentralized social media platforms offer rewards for posting, liking and sharing content which provides a platform to earn. The sooner the social media marketer recognizes the potential of blockchain-based social media, the better they can create strategies for better business and earning and be a part of the new reality. So far so good, here we have explored some social media networks with underlying blockchain technology that are quite popular and disrupting the traditional systems. The significance of blockchain technology is primarily in addressing the major lack of security and data privacy. Amalgamating blockchain technology with social media is promising and the opportunities are huge. It is a decentralized movement where every piece of your data is precisely yours and you earn from your data. A rewarding experience awaits.

References

1. PTT Bilgi Teknolojileri A.S. (2018, June 29). Blockchain@next18 Event. Retrieved from https://www.slideshare.net/obcag/blockchainnext18-event.
2. What is Blockchain Technology. (n.d.). Retrieved from https://www.ibm.com/blockchain/what-is-blockchain.
3. Binance Academy. (2019, November 11). Proof of Work Explained. Retrieved from https://www.binance.vision/blockchain/proof-of-work-explained.
4. 5 Trends Shows How Blockchain Is Changing Social Media. (n.d.). Retrieved from https://hackernoon.com/5-trends-shows-how-blockchain-is-changing-social-media-ba50c975c041.
5. UPADHYAY, N. I. T. I. N. (2019). *Transforming Social Media Business Models through Blockchain*. S.l.: EMERALD GROUP PUBL.
6. Which are the Top 5 Blockchain-Based Social Media Networks? (2018, July 23). Retrieved from https://www.kryptographe.com/top-5-blockchain-based-social-media-networks/.
7. Daniel. (2017, December 12). Sola: Next-Gen Decentralized Social Network Platform: ICO Review. Retrieved from https://www.chipin.com/sola-ico-social-network-user-reward/.
8. Minds. (n.d.). Retrieved from https://www.minds.com/.
9. SoMee.Social. (n.d.). Retrieved from https://somee.social/.
10. Smoke Network Social Blockchain. (n.d.). Retrieved from https://smoke.network/.
11. Microblogging with a Soul (powered by blockchain). (n.d.). Retrieved from https://peepeth.com/about.
12. Mastodon. (n.d.). Retrieved from https://joinmastodon.org/.
13. Platform that Rewards People for Sharing their Lifestyle and Visual Experience. (n.d.). Retrieved from https://steepshot.io/.
14. FAQ. (n.d.). Retrieved from https://about.d.tube/.
15. Blockchain Social Media - Towards User-Controlled Data. (2019, November 5). Retrieved from https://www.leewayhertz.com/blockchain-social-media-platforms/.

Chapter 7

Assessing Security Features of Blockchain Technology

T. Subha

Contents

7.1 Introduction: History of Blockchain

E-cash systems and their design have always been a main research area in the cryptographic field from its inception in the 1980s. But still there is an open space for the verification of data without a trusted third party in e-cash systems. Nakamota initiated Bitcoin in 2009 [1]. This new brand of e-cash has received attention because of the following properties. They are decentralization, unforgeability, and pseudonymity [2]. This attracted remarkable interest to cryptographic research applications.

Bitcoin actually led to the development of blockchain in this technical world. This began after Satoshi Nakamoto's paper which states that the first Bitcoin was mined in 2009. It was an alternative to P2P currency. More varieties of cryptocurrencies started to be introduced to the market soon after the recognition of Bitcoin. Blockchain is the backbone technology for the digital cryptocurrency Bitcoin. Blockchain is termed as a decentralized information-sharing platform for computation. This enables transactions between different authoritative domains that do not trust each other. They collaborate, cooperate, and coordinate with each other [3].

Blockchain is a distributed ledger which is spread among all peers across a network. Each peer in the network holds a copy of this ledger. Each transaction is digitally signed and encrypted and also verified by peers [3]. The ledger information can be seen only by those who are authorized to see.

7.1.1 Working of Blockchain

In this section we explain how blockchain works in detail. Blockchain is a buzzword for many.

The details of how blockchain works are explained below. Imagine if two friends A and B want to transfer/transact money from one account to another account. In a normal scenario, friend A first approaches the bank and asks the bank to transfer the amount to the account of his friend B. This transaction detail is entered into the bank registry that is maintained by the bank. This entry needs to be updated on both the sender's and receiver's accounts. But the problem in this system is that the entries can be easily manipulated or changed by those who are have knowledge about the system. It is easily changeable. This process has been diagrammatically specified in Figure 7.1 [1]. This is the place where blockchain comes into the picture to resolve this issue.

For example, imagine a Google spreadsheet that consists of real-time transactions. This sheet can be shared with multiple users through networks of computers. The users can access this transaction but no one can edit it easily. The spreadsheet is organized as rows and columns and it is treated as blocks in blockchain. A block represents a collection of data in the blockchain. The data get added to the block in

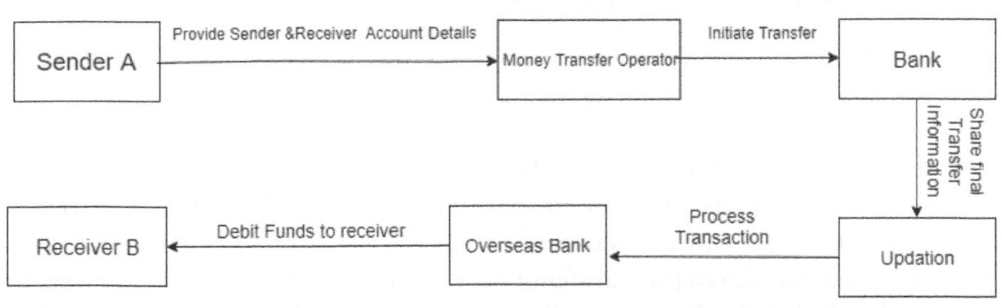

Figure 7.1 Example of traditional transaction in the banking system.

chronological order by creating a chain of blocks that forms a link. The first block in the blockchain is called the genesis block.

Blockchain is developed in such a way that it's called a digital ledger. This digital ledger is duplicated and distributed to thousands of individual computers across the world. These are called nodes, and the interactions between these nodes are periodically updated to the ledger. Each user is given access to a public and private key. These two are the cryptographic keys and are secured one. They allow limited interaction with the system. For example, if there are two users and they agree on an exchange of Bitcoin currency, a particular user can initiate the transaction using their public and private keys. The other user can accept the transaction using their own public and private keys. Both users can submit the transactions to the P2P system.

A unique hash code present in the block checks the transaction information so as to ensure that the transaction agrees with all other prior information. It may deny the transaction if the initiating user does not have the cryptocurrency that they claim for. Otherwise this checking node accepts the transaction and it is added as a new block in the chain.

7.1.1.1 Structure of a Blockchain

The structure of a block is a container that consists of a series of transactions and it is simply a distributed data structure. A block in a blockchain consists of two parts named header and data (transactions).

- Header – this connects all the transactions. If there is a change in any transaction, it will be reflected in a change in the block header.

  The subsequent block headers are connected in a chain. So to make a change requires an update of the entire blockchain.
- Data (transactions)

  Transactions are stored as blocks in the blockchain, and these transactions are hashed and represented as a Merkle hash tree.

7.1.1.2 Steps in Creating a Blockchain

The blockchain structure [2] is represented in Figure 7.2. The working of the blockchain is clearly explained in the following steps.

- **Step 1**: A node creates a transaction by digitally signing it with its own private key. Key pairs are created using a cryptography algorithm. A particular transaction may represent various actions, and it is commonly known as a data structure that shows the transfer of value between different users in a network of the blockchain. It represents information such as logic of transfer of value, source address, destination address, relevant rules, and validation information.

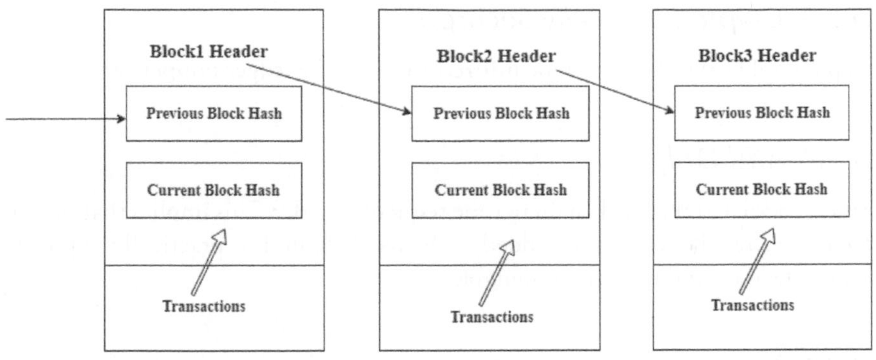

Figure 7.2 Blockchain structure.

- **Step 2**: A transaction is flooded to peers using a Gossip protocol to validate the transaction based on the criteria that are preset. More than one node is required to validate the transaction.
- **Step 3**: A transaction is included in a block after it is validated, and then this information is propagated to the network. The transaction is considered confirmed at this point. All the nodes in a network run proof of work (PoW) and proof of stake (PoS) algorithms to the block to confirm it.
- **Step 4**: Then the newly created block becomes part of the ledger. The next block in the chain links itself automatically to the back of this block. This link is actually known as a hash pointer. This is treated as second confirmation for the transaction. And the block confirms it as a first transaction.
- **Step 5**: Transactions have to be reconfirmed every time a new block is created. To ensure a transaction confirmation is final requires six confirmations usually.

7.1.2 Key Attributes of Blockchain

The following key attributes prove that the usage of blockchain is better than traditional systems [2].

7.1.2.1 Distributed

The ledger is shared among multiple peers in a network. So it is not easy to tamper with or alter the data.

7.1.2.2 Peer to Peer

There is no central authority or central control to manipulate or monitor the transactions. Every participant that participates in a transaction talks to the others directly. This makes data exchange easy with the involvement of third parties.

7.1.2.3 Cryptographically Secured

Cryptographic algorithms can be utilized to make the ledger tamper-proof.

7.1.2.4 Add Only

Data are added to the blockchain in time sequential order. This implies that the data cannot be altered once they are added to the blockchain. It is practically impossible to alter the data and they are immutable.

7.1.2.5 Consensus

This is the most critical attribute of all the attributes of the blockchain. The ledger data can be updated via consensus [4]. It leads to the property of decentralization, i.e. no central authority is involved in updating the ledger. So in the case of any update to the ledger, it must strictly follow the protocol designed and the update should be validated. Finally, it is added to the blockchain only if all the participating peers and nodes reach consensus in a network.

7.1.3 Types of Blockchain

Mainly there are three different types of blockchain are present [5]. They are,

- Public blockchain
- Private blockchain
- Consortium/federated blockchain
- Hybrid blockchain

These types are represented diagrammatically in Figure 7.3 [2]. These types and its uses are clearly explained below.

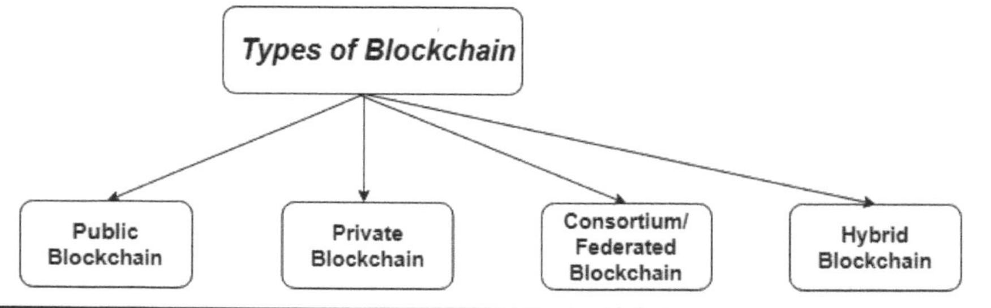

Figure 7.3 Blockchain types.

7.1.3.1 Public Blockchain

It is the blockchain for the public and it has no restrictions. Anyone with an Internet connection is allowed to send a transaction (i.e. reading/writing/auditing) and can become a validator in the group. This type of blockchain is transparent and open type. If this is going to be the case, then who is responsible for transaction confirmation? Because anyone in the group is allowed to review anything at a given point of time. So this is done with the help of decentralized consensus mechanisms such as PoW and PoS; all these nodes are participating in the execution of consensus.

This blockchain is *"for the people, by the people, and of the people"*.

The largest known public blockchain examples are Bitcoin, Litecoin, and Ethereum.

7.1.3.2 Private Blockchain

Private blockchains are for individuals or organizations. One cannot join in a private blockchain unless invited by the network administrators. Restrictions are put on participant access and validator access. This type of blockchain is convenient for users who do not want to share their sensitive data with a public blockchain. It is mainly used for accounting and for keeping records without compromising autonomy. In a private blockchain, consensus is granted by a central authority and he is responsible for granting mining access to all or not to anyone. Again this comes under a centralized network which contradicts the idea of blockchain, but it is cryptographically secured.

Bankchain is an example of a private blockchain. Private blockchains are also called permissioned.

7.1.3.3 Consortium/Federated Blockchain

A consortium blockchain is a type of semi-decentralized blockchain but it is permissioned too. Multiple companies or multiple individuals combine together to operate the node in a network. These groups are known as a consortium or federation, and their combined nature helps in making decisions for the benefit of the entire network. Administrators will grant the reading request and consensus request to a limited set of trusted nodes who can take part in consensus execution. It is much faster in accomplishing things, and there is no single point of failure. For example, the world's top 20 financial institutions are grouped under one consortium. At least 15 companies should vote for or validate the transaction in order to add one block or to make decisions in this type.

R3 and EMF are examples of consortium blockchains.

7.1.3.4 Hybrid Blockchain

As the name suggests it's a combination of different characteristics of public and private blockchains. With this type we can decide which information is made public and which information is private.

7.1.4 Pillars of Blockchain Technology

The following three are the important pillars of blockchain technology [3].

7.1.4.1 Decentralization

Users or industries know very well about centralized servers, and they have been working on it since before Bitcoin and BitTorrent came into the picture. The idea of storing data and retrieving the data from a central entity is an easy one. Users have to interact only with the central authority for the required information. Banks are one of the well-known examples for centralized systems.

A bank stores your money in an account and maintains all the transactions in a centralized server. If you want to transfer money into your friend's account it must be done through the bank only. We can look at the client–server example as the best one for a centralized system. For example, if you (client) search something in a Google search engine, it queries the server (server) for the result and returns it back to you. This is another good example for a client–server system. We have been working with centralized systems for many years and at the same time they also have vulnerabilities.

- As data are stored in a single central repository, this one spot becomes an easy target for hackers and others.
- The entire system halts in case of any upgrade to be carried out in a centralized system.
- If the system is shut down for unknown reasons, then there is no way to retrieve the data from a centralized system.
- What happens to the data if they get corrupted or become malicious in the worst case? The data will be completely compromised in that case.

If we try to take this centralized entity out, it leads to the concept of decentralization. In decentralization the data are not stored or maintained by a single entity. In fact, they are owned by every user participating in the network. If two users want to communicate in a blockchain then they can directly communicate without the involvement of a third party. Bitcoin is implemented using this ideology only. You are the only one responsible for your money, and you can transfer it to anyone without going through the bank.

7.1.4.2 Immutability

Immutability is nothing but that the data cannot be tampered with once they are entered into the blockchain. Blockchain is able to achieve this unique property with the help of cryptography hash value. Hashing is the process of producing a fixed-length hash as output, and it takes an input string of any length. Transactions are the inputs in Bitcoins, and they run through a cryptography hash algorithm such as SHA-256 (used by Bitcoin) and it produces fixed-length output.

Example: SHA-256 hash function

Even if the input is of an arbitrary length it is easy to remember the 256 bits of hash value output. The cryptography hash function has many attractive features. One of the important properties that we need to keep in mind is the "Avalanche Effect". If you try to make even a small change in the input value, it will be reflected in a huge change in the output hash value.

Blockchain is organized as a similar structure of linked lists which contain data and a hash pointer that points to the previous block. A hash pointer is a kind of pointer which points the hash value of data, i.e. present inside the previous block. This property makes blockchain immutable and amazingly reliable. If anyone tries to change a single bit of data in block 4, it is reflected in a huge change in the data stored in block 3 and in turn it affects the data stored in block 2 and it continues further. The change in hash value is used to find whether the transactions are altered by any means. But it is completely not possible based on the property of hash.

7.1.4.3 Transparency

The most attractive feature of blockchain technology is the degree of transparency it can provide in a network. Another important aspect of blockchain technology is the degree of privacy that it can provide. Transparency is basically concerned with the quality of being clear and understandable without ambiguity. What happens actually is blockchain hides the user identity by using complex cryptography algorithms. So while looking for a particular transaction record it does not look like "alice sent bob 2btc" in the original record. Instead it will look like encrypted content "2BM2hdskjfhdfgRvndkERFVbjb237bu sent 2 btc".

The transactions in Ethereum blockchain are still seen by the public address even if the real identity is hidden. So blockchain ensures privacy, and this kind of transparency has never existed in any of the earlier financial systems. It can be checked easily in Internet Explorer if we know the public address of big companies that do transactions using their cryptocurrency. But the big financial companies won't perform all their transactions using cryptocurrency. The big boom in blockchain is because it can be integrated in supply chain management.

7.1.5 Benefits of Blockchain

The following lists out the benefits of blockchain in detail.

- No central control or no central authority is involved.
- Eliminates additional costs if any trust is built between blockchain members.
- Transactions are signed digitally using asset owner's private/public key pair.
- Data cannot be altered easily once they are recorded and added as a block.
- Distributed ledgers efficiently store transactions between members in a verifiable and permanent way.
- Transactions do not have to be data; they can be a code or smart contract in blockchain.

7.2 Literature Review

A blockchain is a shared and distributed ledger, in simple terms it is a chain of blocks. The block talks about digital information and this information is stored in a public database is called chain. The appealing aspect of blockchain technology is easy to track the resources without having a centralized trusted authority [6]. It permits two different parties to communicate and exchange resources easily wherein the distributed decisions are taken by the majority rather than by a single centralized entity. It provides security from attackers who try to compromise centralized systems or controllers.

Resources can be of different types such as tangible or intangible resources. Money, cars, houses, and land are classified as tangible resources, and copyrights, IPR, and digital documents are classified as intangible resources. As of late, the blockchain innovation has pulled in huge enthusiasm from both the scholarly community and industry. The innovation began with Bitcoin, a cryptographic money, also called cryptocurrency, that has arrived at a capitalization of 180 billion dollars as of January 2018 [7, 8].

As indicated by the Gartner report in 2016, the blockchain innovation is attracting billions of dollars in research and considerably more innovation is relied upon to come sooner rather than later [9]. The innovation as of now traverses a few applications that are well-known and driving the systems administrators to explore further innovations in the recent domains such as supply chain, retail etc. Such applications incorporate medicinal services [10], the Internet of Things (IoT), and cloud storage [11]. The innovations of blockchain technology pave the way to the development of new platforms and applications. Many researches have written several survey papers to highlight the benefits of this technology in current areas of application. Such examples include blockchain technology for healthcare [10], IoT [12], blockchain adoption in higher education [13], blockchain as a service platform [14], decentralized digital currencies [15], and blockchain-enabled smart contracts [16].

The blockchain innovations and its potential have been demonstrated in many of the popular applications. As of now regular monitoring and security services comprising confidentiality, authentication, integrity, privacy, and provenance are offered by third-party services or brokers. They use inefficient distributed applications to provide these services to customers. As a result of this it is found that the security is the main threat for these applications. Researchers can focus on the above-mentioned areas to give insight about the use or applications of blockchain to leverage these security issues.

7.2.1 Applications of Blockchain Technology

Bitcoin was the first application of blockchain. It's a kind of cryptocurrency or digital currency based on blockchain technology. Its primary use is for making transactions online using the Internet as we do in the real world. Now blockchain is being adopted by several applications because of the success of Bitcoin. Specialized applications are financial, marketing, supply chain, management, healthcare, IoT, and manufacturing. At the same time cyber criminals are getting opportunities to get involved in cyber-crime. Most of the attacks, for example 51% of the attacks, are security-based attacks on Bitcoin in which the hacker tries to take control of the system, its working mechanism, etc.

Blockchain technology can be applied in wider areas for various applications. The sectors are agriculture, IoT, business, food processing, finance, healthcare, manufacturing, and various other sectors. Table 7.1 describes applications of blockchain technology in different sectors.

7.3 Security Features in Blockchain

We may question why blockchain is a secure system even if the ledger concept is not new to our system. But the interactions about security parameters open out in a digital environment and may bring vulnerability to blockchain technology in terms of a security threat.

Is blockchain secure? Security becomes a big question mark when our personal data are stored in an online repository. Blockchain is inherently secure. It applies powerful cryptography techniques to the individuals who take part in blockchain transactions so they can hold ownership of an address, crypto assets of public and private keys associated with it. The passwords are generated using random combination of letters and numbers called alphanumeric characters. This solves issues such as identity theft because addresses are not associated with the owner's identity directly. Private keys are considerably larger and more secure. Blockchain offers a greater level of security for individuals as it removes the need for weak and easily compromised passwords.

Table 7.1 Blockchain Technology in Various Applications

Agriculture	Agriculture and soil-related data processing, sales, marketing, and shipping of agro products, yields, etc.
Distribution	Transportation of sales record, marketing record, digital currency, mining chip, used goods for sale
Business	Importing and exporting data and digital records by software industries, processing of transaction data, all other data which have finance value
Energy	Data related to energy generation, data related to raw material, availability of resources, supplier-related data, demand data records, maintenance of tariff data, on-demand supply, resource tracking
Food	Delivery and shipping of food data records, food packing details, online ordering and transaction data, quality assurance data
Finance	Money deposit and transfer details, smart contract and security details, social banking, digital transaction data, cryptocurrency
Manufacturing	Product assurance and guarantee information, product warranty information, robotics, manufacturing and production data, supplier components, tracking of raw material
Healthcare	Electronic records, automating hospital process, billing, information stored at hospital server, costs of healthcare
Smart city	Water management, energy management, pollution control data, digital data and transaction details, smart data maintenance, smart transaction
Transport and logistics	Logistics services, shipping details and delivery data, maintenance of toll data, vehicle tracking, tracking of shipment containers
Others	Data and details with respect to economy, digital data, artwork, ownership, jewelry design, space development, government, voting, etc.

7.3.1 Issues in Blockchain Security

Blockchain is an increasingly growing and popular technology. It is also simultaneously queried by different leading industries, since it is an adaptable technology and can be utilized by many industries [17].

This chapter discusses the security and privacy issues in blockchain in detail. Security in blockchain is needed to protect the data and the transaction information against internal, external, and unintentional threats and peripheral attacks. Typically, this helps to detect, prevent, and provide appropriate responses to the threats based on security policies, rules, tools, and IT services.

7.3.1.1 Defense in Penetration

A unique strategy is applied to protect the data in defense-related applications using infinite counter measures. The principle of protecting data in multiple layers is enforced in blockchain rather than providing security to a single layer.

7.3.1.2 Manage Vulnerabilities

Vulnerability checking involves identification, authentication, modification, and patching of them whenever is required.

7.3.1.3 Minimum Privilege

In this principle access to data is restricted to the lowest level possible to promote a high level of security.

7.3.1.4 Manage Risk

The identification of risk, assessing risk, and control of risks are processed in this phase.

7.3.1.5 Manage Patches

Required patches are installed to patch the damaged part such as code, operating system, firmware, and applications.

Many techniques have been used in blockchain to achieve security for the transaction data or block data irrespective of the usage. Bitcoin applications use encryption techniques for providing safety to the data. The combinations of public/private key pairs are used to encrypt and decrypt data securely. The most secure part of the blockchain is the longest one which is considered the authentic one. Fifty-one percent of the major attacks and fork problems have been reduced by means of blockchain. By considering the longest chain as the authentic one in blockchain, it makes the other attacks null and void.

7.3.2 Privacy of a Blockchain

Privacy is an important parameter for users who maintain data in online repositories. This is the individual rights of a person to safeguard their data. It is possible to perform transactions without leaking identical information in blockchain [18]. So it helps to achieve privacy. The user is allowed to perform their work without showcasing it to the entire network. The primary goal of enhancing privacy in blockchain is to make it harder to copy other users' crypto profiles. Several volumes of variations are perceived while applying blockchain in privacy.

7.3.2.1 Significant Characteristics of Blockchain Privacy

- **Stored data** – blockchain provides a way to store all forms of data. The privacy level of data for individual and organizational data varies with different owners. Even if privacy rules are applied to personal data, even more stringent rules are applied to organizational sensitive data.
- **Storage distribution** – full nodes are nodes in the blockchain that store complete copies of information about the blockchain. Full nodes with the combination of an append-only characteristic lead to data redundancy. This data redundancy supports major key features of the blockchain, namely transparency and verifiability. The access levels of these two key features are decided based on the application compatibility with data minimization.
- **Append only** – it is very difficult to change the data of the previous block without being detected. Sometimes this property will not meet their purpose if the data is recorded incorrectly. So special attention and care are required in assigning rights to data.
- **Private vs public blockchain** – accessibility is the remarkable parameter from the point of security and privacy. Further, the data can be encrypted to be made available for authorized users as everyone in the network has the copy of the data.
- **Permissioned vs non-permissioned blockchain types** – users in public and non-permissioned blockchain types are permitted to add data.

7.3.3 Blockchain Applications Security, Privacy Challenges, and Solutions

The following section explores security and privacy challenges, and their solutions with regard to blockchain applications [19].

7.3.3.1 Blockchain in Healthcare

Blockchain contributions to the healthcare domain need to maintain data privacy, need to be public and secure, and also should support scalability at any point in time. The blocks present in a healthcare blockchain represent data, images, and documents about health records. Blockchain in healthcare faces the data storage problem and limitations on throughput. If Bitcoin is chosen as a model for storing data, then every individual in a network will contain a copy of the healthcare data which may lead to security problems. This is not the best storage method, and bandwidth severity may occur. Network resources are wasted in this method. If blockchain needs to be implemented in healthcare, an access control manager should be added to specify the access and management of data, and storage of data.

In reality the blockchain maintains an index or list of all users' data. This acts as a catalog that stores metadata about the patient's record and the location of data where it is being stored. These data are accessed by authorized users. Data efficiency can be improved by further encrypting the data, adding a timestamp to the data, and retrieving the data by assigning a unique identifier to each record. A data lake is defined as a blockchain data repository where all healthcare data are stored in blockchain. They are extremely valuable data as they contain any form of data. It provide all kinds of technical support such as querying the result, analyzing the text, mining the data, and integration with machine learning [4]. The complete view of blockchain transactions in a healthcare domain is represented in Figure 7.4.

7.3.3.2 Blockchain in Finance

Bitcoin is a popular digital currency which led to the development of blockchain later. Further, blockchain is used in many cryptocurrencies, namely Ethereum, Peercoin, Altercoin, Karma, Hashcode, and Binarycoin.

Bitcoin is the primary structure for most of the digital cryptocurrencies, but they use different consensus algorithms for the verification and validation of data.

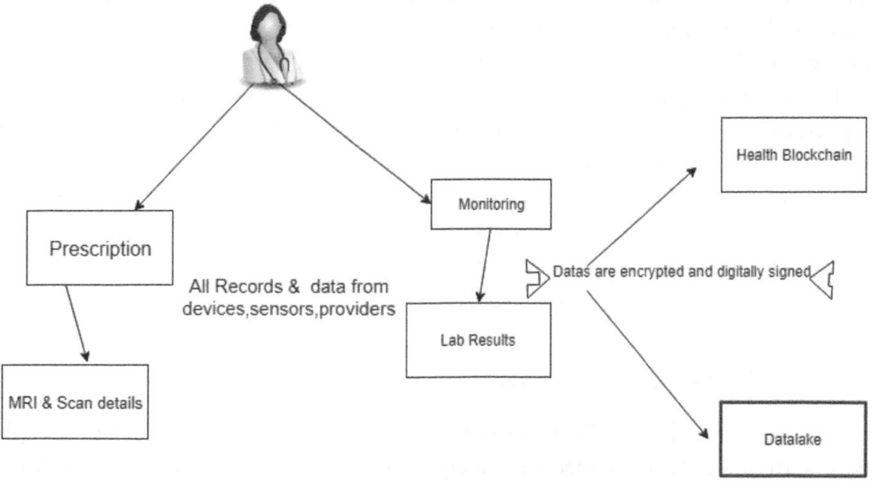

Figure 7.4 Blockchain in healthcare application.

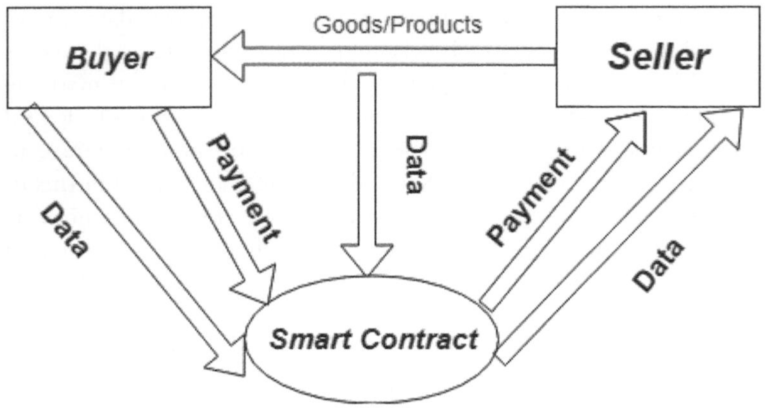

Figure 7.5 Blockchain in finance application.

The primary part of a blockchain in the financial domain is the smart contract. A detailed look at this is presented in the Figure 7.5. We will see the security and privacy issues in finance domain blockchains. The common security problem in implementing blockchain in any organization is they must ensure that data are accessed by only authorized users. Checking the data access by authorized users is mandatory in blockchain fundamentals. It is necessary to enforce the implementation of authentication and authorization control [20]. The importance of blockchain in the finance industry is diagrammatically given in Figure 7.5.

Blockchain also ensures one of the basic security parameters, confidentiality, by allowing data blocks to be encrypted. Different layers of security can be implemented with the help of a combination of private and public key pairs. Data integrity and consistency are preserved. Immutability and traceability ensure the integrity of data in blockchain. This is another important parameter in providing security. In addition, hash values ensure that no blocks of data can be altered or copied.

Ensuring the privacy of individual user data or organizational data is another important aspect of security in blockchain. There are many ways to achieve this property. One is to encrypt the personal information of every user. If a key is also forgotten, then also it ensures privacy and security, since no one can access data if they don't possess the key. This is because all the transactions are timestamped and signed using digital signature algorithm which ensures non-repudiation. So it increases the reliability of blockchain in the finance domain.

7.3.3.3 Blockchain in Internet of Things (IoT)

IoT can be defined as an interconnection of computing systems, devices, people, sensors, digital machines, mechanical equipment, objects, etc., in a network where they are able to communicate with each other without human-to-human or human-to-machine interaction [20]. All the devices participating in IoT are assigned with unique identifiers. The primary requirement of IoT in blockchain is the storing of data in edge devices and access of those data by users. At any point in time the user may want to access the data stored in remote locations securely and the different ways in which it wants to ensure data privacy.

The user has a provision to set up the password and required access control while creating a user account itself. The user sends data for storing after checking required permission controls and by extracting the previous block number and hash value. Storage availability is confirmed after validating the transaction. In some applications the service provider wants to access the data. In that case a request is sent to the cluster head by the service provider and service requester. Then it is verified by the cluster head with its members or with another cluster head. Safe answer method and noise introduction methods are followed for protecting data and ensuring user privacy.

Finally, whether the user has accessed the data or not is set by the multisig transaction value to 1 or 0. Multisig transactions are treated as a proof by storing the date that was sent by the user. Any misconduct can be easily intimated to other users because of this multisig. The advantage of using IoT in blockchain is that trust is built between communicating parties, reducing cost and speeding up transactions.

Blockchain in IoT helps business growth and accelerates its importance. The security concern of IoT is sensing data, processing and storing of data, and communicating data finally. With the help of public blockchain this can be implemented easily as everyone in the network uses his private key to safeguard it. A blockchain-based IoT model builds trust since there is no central authority involved [12]. The biggest challenge that needs to be addressed in IoT blockchain models is scalability, because the number of requests to access data will be high if the amount of computing devices increases day by day.

7.3.3.4 Blockchain in Mobile Applications

Mobile applications are a special kind of applications that are designed to be suitable for mobile phones and tablets. Blockchain is able to provide peer-to-peer support for direct payment and peer-to-peer file transfer in mobile applications. But it does not support speed games which miners use to validate a particular transaction. The edge computing concept can be utilized to implement blockchain in mobile processing.

Consider there are N mobile application users such as N = 1, 2, 3 … N. Now every user is trying to solve the puzzle using the hashing concept to get rewards. Edge computing devices can be deployed with service providers. Mobile users run mobile blockchain applications that support edge computing for miners. The service provider charges for the amount of service utilized by the mobile users. Blockchain can be applied to many different mobile applications from a security point of view since it has no single point of failure [20]. Figure 7.6 explains how the transactions are carried out in mobile applications [19].

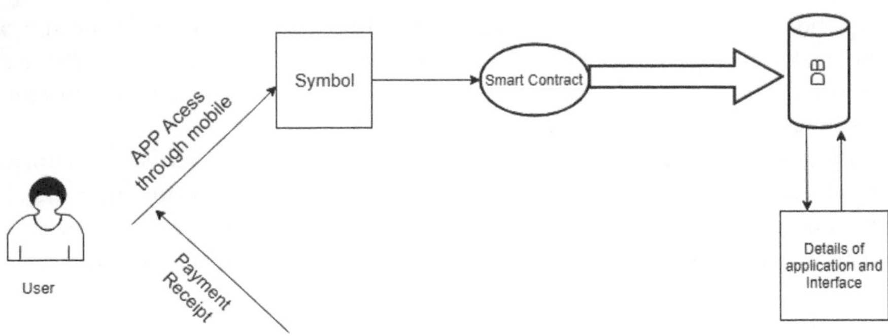

Figure 7.6 Blockchain in mobile applications.

It is recommended to use blockchain in mobile applications where a strict requirement of authentication is required for the protection of data. The major use of mobile applications is in accessing digital wallets stored in mobile phones. Users are allowed to pay for their transactions through mobile devices as all the payments are supported by mobile applications using their digital wallets. This is accomplished by edge computing services provided by the service provider. So it makes the transaction portable and convenient.

7.3.3.5 Blockchain in Defense

Blockchain will play an important role in defense in the near future because the defense applications are cyber-enabled systems and they must provide utmost security to the data they handle. The current cyber systems lack the technology to address the growing cyber threat. Blockchain helps to reduce the error rate in cyber defense systems by providing transparency, fault tolerance, and trust in transactions. The following security features of blockchain such as hashing, back linked list of data structure, consensus algorithm, and immutability play an important role in implementing blockchain applications [21].

Cyber defense applications require their data to be confidential and accessed only by authorized users. Blockchain accomplishes authentication by granting access only to the permissioned users. Confidentiality of the data can be ensured by encrypting the data before storing it in blockchain. Then access control can be defined at various levels to restrict access to the data. Blockchain is capable enough of playing a operational or support role in cyber defense applications. They are as follows.

7.3.3.5.1 Cyber Defense

In cyber defense applications blockchain transmits data to all other nodes in a network and applies a consensus algorithm to verify and validate the transactions. The data cannot be manipulated once they are time stamped and stored as a block. In case of a data update by the authorized users, then it needs to be newly time stamped and maintain proper log information. The weapon, components, and its details can be imaged, hashed, and then it is stored in a secure data base. They can be monitored continuously using blockchain applications [22].

7.3.3.5.2 Supply Chain Management

The concern of supply chain management in blockchain applications is the need for owner traceability and to maintain the origin of data. Blockchain is able to give a solution to this.

7.3.3.5.3 Resilient Communications

The secure communication of blockchain helps to provide resilient communication in contested environments. This helps to make the data transfer reliable between the nodes across the world.

7.3.3.6 Blockchain in Automobile Industry

Nowadays the vehicles that are moving on the roads are connected online and able to identify the traffic patterns, locations, and other details. Blockchain can play an important role in achieving this. Some security features need to be addressed for smart vehicles. They are as follows.

7.3.3.6.1 Scalability

This system must be scalable based on demand because in VANET the numerous vehicles with onboard electronics equipment get added to the network frequently.

7.3.3.6.2 Safety

It should not generate new security threats, and the malfunctioning of devices leads to false driving behavior. So a blockchain implemented in smart vehicle architecture needs to protect users from security threats.

7.3.3.6.3 Decentralization

A decentralized architecture is required for storing the data since data stored in a centralized repository are prone to a single point of failure. Blockchain offers this decentralized concept which is suitable for incorporating into automobiles.

7.3.3.6.4 Maintenance

The architecture of smart vehicles is comprised of a lot of hardware and software equipment. So blockchain should support this maintainability for a certain period of time and should have the option for maintainability.

The main critical challenge in the automobile industry is wireless remote software updates (WRSU). WRSU fixes the errors in the control unit whenever a software upgrade is needed, so that the complete life cycle of a vehicle is clearly captured. The focus is required in the WRSU security architecture for the better management of vehicular ad-hoc network [20].

The software update process of WSRU architecture is illustrated in the following steps.

- ***Step 1***: Software provider initiates software upgrade if the latest version is released, then it is stored in the cloud that is made available to the overlay nodes.
- ***Step 2***: User creates a multisig transaction and it is encrypted with public/private key pair and signed digitally.
- ***Step 3***: Car manufacturers and block managers use their public key to forward the transactions with the list of all the keys through which it achieves integrity.
- ***Step 4***: Transaction is further sent to overlay since it is not treated as a valid transaction by all block managers because the current transaction contains only a single key.
- ***Step 5***: Block managers broadcast transaction to the network. After receiving the transaction, the block manager of the cluster verifies the software update and acknowledges it.
- ***Step 6***: This transaction is further broadcast to all block managers. Further block managers verify the software update with the public key of the car manufacturer and software provider.
- ***Step 7***: The smart vehicle verifies the transaction received from the block managers.
- ***Step 8***: Finally, the vehicle can download the software directly from cloud storage by using their own authentication parameter.

  Blockchain can also be implemented in various domains of automotive industry such as finance, insurance, leasing of car, electric and smart charging services, etc.

7.3.4 Challenges in Blockchain

This section describes the challenges faced by blockchain technology.

7.3.4.1 Privacy Leakage

The problem in blockchain privacy is the leakage of transactional privacy data because the details and the balances of public keys are visible to all in the network. Anonymity in blockchain is classified into mixing solutions and anonymous solutions. Mixing is a kind of service that offers anonymity by transferring funds from multiple inputs addressed to multiple output addresses. Anonymous is another type of service which unlinks the payment origin for a particular transaction. It stops the intruder from analyzing the transactional data.

7.3.4.2 Scalability

Blockchain needs to be scalable because of its increasing volume of applications. Every node should store the copy of the transaction to validate the data. First the transaction source needs to be validated before the transaction is validated. Block size also plays a role in the scalability issue because miners want to validate bigger transactions in order to receive higher transactional fees. The scalability issue may be categorized into storage optimization and redesigning of blockchains. Solutions are invited from research community to address these challenges.

7.3.4.3 Personal Identifiable Information

PII information can be used to extract an individual's identity. PII can be addressed with respect to location and communication privacy.

7.3.4.4 Selfish Mining

Another challenge faced by blockchain is selfish mining. A block is called a susceptible block if a small amount of hashing power is used. In this method, miners do not broadcast the mined block's result immediately. They open a private branch in which the mined blocks are posted after certain conditions are met. True miners waste their resources and a lot of time in mining the block whereas the private link is mined by selfish miners.

7.3.4.5 Security

Security is the main important feature in any upcoming new-generation technology. Security can be detailed in terms of confidentiality, integrity, and availability. This always remains an open challenge in public blockchain. Confidentiality is poor in distributed systems since it copies information to entire network. The integrity principle of CIA triad still faces many more challenges in real time; even integrity of information proves to be not altered. Read availability is high compared to write availability in blockchain.

7.3.5 Popular Blockchain Use Cases

The following discusses the popular use cases in blockchain.

7.3.5.1 Cryptocurrency

Risk linked with cryptocurrency: As digital currencies are encrypted it is possible only to identify currency and not its owner. The currency is owned by whoever owns the encryption key. So if a currency is stolen, the money is never recoverable. There is no way to get it back.

Solution: Any method to store the encryption keys in a secure vault of trust.

7.3.5.2 Smart Contracts

Risk linked with smart contracts: A smart contract is a computer-based program that details their capability to self-execute and enforce the contract rules. The smart contract rules and self-execution are altered if the blockchain is breached. It breaks the basic trust of a blockchain and it removes a way of doing business between parties without the involvement of a middleman.

Solution: Secure the terms and self-execute factors of a smart contract and store the encryption keys in a hardware-based root of trust with an anonymous party with secured authentication. That ensures that no one can access the data.

7.3.5.3 Internet of Things (IoT)

Risk linked with IoT: The data are placed and operated in a central repository in the Internet of Things that makes the system more vulnerable. Recently Mirai-style botnets allowed the hackers to take control of the 100 IoT devices connected in a network and access to its information due to less security control. IoT devices are basically protected with default passwords are easily prone to distributed-denial-of-service (DDoS) attacks that are launched by hackers.

Solution: A distributed trust model in blockchain helps to protect the IoT from DDoS attacks. It removes single point of failure and enables device networks to protect themselves by means of other ways. For example, any node in the network may quarantine the node that behaves unusually.

7.4 Conclusion

This chapter clearly discusses the blockchain technology, its properties, and how blockchain can be implemented in the various domains of applications such as IoT, healthcare, supply chain management, automobiles, finance, etc. Whenever the new technology is released it also brings security threats simultaneously. Security issues in blockchain have been discussed in detail. Popular use cases of blockchain technology have been explained briefly.

References

1. https://blog.goodaudience.com/blockchain-for-beginners-what-is-blockchain/.
2. https://cointelegraph.com/bitcoin-for-beginners/how-blockchain-technology-works-guide-for-beginners/.
3. https://blockgeeks.com/guides/what-is-blockchain-technology/.
4. Z. Zheng et al. 2017. "An Overview of Blockchain Technology: Architecture, Consensus, and Future Trends". *Proceedings of the 2017 IEEE BigData Congress*, Honolulu, Hawaii, pp. 557–564.
5. S. S. N. L. Priyanka and A. Nagaratnam. 2018. "Blockchain Evolution – A Survey Paper". *IJSRSET* 4(8): ISSN:2395-1990.
6. Tara Salman et al. 2018. "Security Services Using Blockchains: A State of the Art Survey". *IEEE Communications Surveys & Tutorials* 21(1): 858–880.
7. I. Eyal et al. 2016. "Bitcoin-NG: A Scalable Blockchain Protocol". In *Proceedings of 13th Usenix Conf. Network System Design and Implementation (NSDI)*, Berkeley, CA, pp. 45–59.
8. Crypto Currency Market Capitalization. Accessed August15, 2017. [Online]. Available: https://coinmarketcap.com/currencies/.
9. STAMFORD. Gartnet's 2016 Hype Cycle for Emerging Technologies Maps the Journey to Digital Business, August, 2016. [online]. Available: http//www.gartner.com/newsroom/id/3412017.
10. M. Mettler. 2016. "Blockchain Technology in Healthcare: The Revolution Starts Here". In *Proceedings of IEEE 18th International Conference on e-Health Networks Applications Services(Healthcom)*, Munich, Germany, pp. 1–3.
11. K. Christidis and M. Devetsikiotis. 2016. "Blockchains and Smart contracts for the Internet of Things". *IEEE Access* 4: 2292–2303.
12. M. Conoscenti, A. Vetro and J. C. De Martin. 2016. "Blockchain for the Internet of Things: A Systematic Literature Review". In *Proceedings of IEEE/ACS 13th International Conference on Computer Systems Applications (AICCSA)*, Agadir, Morocco, pp. 1–6.
13. Khoula Al Harthy et al. 2019. "The Upcoming Blockchain Adoption in Higher Education: Requirements and Process". *IEEE International Conference.*
14. W. Zheng et al. 2019. "NutBaaS: A Blockchain-as-a-service Platform". *IEEE Access.*
15. S. Ahamad et al. 2013. "A Survey on Crypto Currencies". In *Proceedings of 4th International Conference on Advanced Computer Science (AETACS)*, pp. 42–48.
16. S. Wang et al. 2019. "Blockchain Enabled Smart Contracts: Architecture, Applications, and Future Trends". *IEEE Transactions on Systems, Man, and Cybernetics: Systems* 49(11): 2266–2277.
17. J. Yli-Huumo et al. 2016. "Where is Current Research on Blockchain Technology? – A Systematic Review". *PLoS ONE* 11(10): e0163477. doi.10.1371/Journal, 2016.
18. G. Zyskind et al. 2015. "Decentralizing Privacy: Using blockchain to protect personal data". *Security and Privacy Workshops (SPW), 2015 IEEE*, IEEE, pp.180–184.
19. Zibin Zheng et al. 2018. "Blockchain Challenges and Opportunities: A Survey". *International Journal of Web and Grid Services* 14(4): 352–375.
20. Archana Prashanth Joshi, Meng Han and Yan Wang. 2018. "A Survey on Security and Privacy Issues of Blockchain Technology". *Mathematical Foundations of Computing* 1(2): 121–147.

21. J. Mendling et al. 2018. "Blockchains for Business Process Management – Challenges and Opportunities". *ACM Transactions on Management Information Systems (TMIS)* 9(1): 1–16. Article No.4.
22. W. Tirenin and D. Faatz. 1999. "A Concept for Strategic Cyber Defense". In *Military Communications Conference Proceedings, MILCOM 1999*, IEEE, vol. 1, pp. 458–463.

Chapter 8

Merger of Artificial Intelligence and Blockchain

Pooja Saigal

Contents

8.1 Introduction

Artificial intelligence and blockchain are the two most promising technologies that have evolved in the last few years. In a very short span of time they have permeated into almost every possible industry. AI enables machines to make decisions like human beings and can handle complex problems, whereas blockchain provides a distributed data environment and offers data transparency, security and privacy. Although these technologies look quite different from each other, there are prospects to combine them. This amalgamation would take the strengths of both AI and blockchain, so that the resulting technology would be robust and more efficient than either of them individually. These two technologies complement each other and can therefore reduce the weaknesses of each other. Since AI requires huge amounts of data for making predictions, these datasets can be stored on a distributed platform provided by blockchain. Blockchain could be benefited by the use of AI to create autonomous organizations or to monetize user-controlled data. Both AI and blockchain are in their nascent stage and are going to evolve individually. In this chapter, we'll see how these two technologies interact and what the future prospects for their merger are.

8.2 Artificial Intelligence

Human beings have the ability to imagine and convert it to reality. Its true manifestation is realized in the last few decades. Although a common man may not realize it, artificial intelligence (AI) is already embedded into our life. While searching for keywords on the Internet, on our way to work, purchasing online or while chatting with friends, we are most likely using AI in the background. The airline industry has been using AI for decades now, to regulate air traffic and for efficient landing and take-off schedules. Boeing planes involve human steering mostly at the time of landing and take-off, whereas for the rest of the flight, autopilot takes command. Google Maps uses location data from smartphones and analyzes the traffic patterns for that location in real time. By using feedback data of visitors and user-reported cases like breakdown, construction, accidents, etc., Maps can provide more accurate information and route suggestions. These suggestions significantly reduce the commuting time by informing the user about the fastest route between two locations. Various cab booking apps like Ola, Uber, etc., use AI to determine the cost of the ride, best route and for booking nearby cabs on request with minimum

Figure 8.1 Applications of artificial intelligence.

arrival time. Their ride-sharing feature optimally books the passengers to minimize detours. They use AI to determine surge pricing, estimated time of arrival (ETAs), meal delivery time and also for fraud detection. Figure 8.1 shows a few popular applications of AI.

AI plays a major role in managing social networking sites like Facebook, LinkedIn, and Pinterest. Facebook has an AI-based feature called tagging that can identify faces in any uploaded image and can suggest names for them. It is based on a face recognition algorithm of machine learning (ML). ML is a branch of AI that develops algorithms which can learn from data and can predict outcomes for unknown data. For a common man, it is not easy to understand how Facebook is able to recognize his family members and friends. For tagging, Facebook uses an ML algorithm built upon artificial neural networks (ANN) that mimics the human brain to recognize human faces. In 2016, Facebook introduced DeepText [1] which is a text-understanding engine that can understand the textual content of thousands of posts, from multiple languages (around 20), with accuracy close to human understanding. DeepText helps to identify and mark the content that is most relevant. For example, a property broker is most interested in posts related to sale and purchase. So, DeepText can identify such posts and displays the most relevant content. Similarly, DeepText helps popular public figures to automatically identify the most relevant text from the comments on their posts. Facebook acquired Instagram which uses machine learning to suggest emojis. Recently, emojis have replaced text and are used to show sentiments. The ML algorithm also tries to understand user sentiments, by studying the type of emojis used by users at a given point of time. Instagram has made emojis very popular not only among the young generation, but for almost every user of any age. It has totally changed the way people communicate with each other. Only a decade before, no one would have imagined that simple emoji graphics would be used to show the sentiments of a distant person. There is another popular social media app called Pinterest that has a huge image and video database. Pinterest uses computer vision and pattern recognition techniques of ML to automate the process of object identification in images and videos. Pinterest also recommends similar pins to the users. The visual search feature of Pinterest is based on image matching and retrieves the best matches for any given image. Snapchat is very popular these days due to its feature called Lenses. These are filters that can track facial movements and can add animated effects to the images. Snapchat presents a variety of filters and keeps updating them on a regular basis. These filters use ML algorithms to learn the movements of faces in real-time videos.

Email inboxes use an AI-enabled powerful feature for spam filtering. It is not based on a fixed set of words, but it continuously learns from the messages, related metadata and user responses. Google uses an AI approach to mark emails as important based on the sender's data or message content and categorizes emails under primary, forum, social, promotion and update inboxes. Gmail also learns if a user marks an email as spam or important. Recently, many banks have started to provide the facility to deposit cheques through a smartphone app. Banks rely on companies like Mitek [2] that use identity verification and mobile capture technology, built on artificial intelligence algorithms, to decipher and convert handwriting on cheques. Banks also use AI to find fraudulent transactions, by creating systems which can learn from fraud transaction data. Atiya [3] proposed neural network algorithms to determine the credit worthiness of a loan applicant, based on historical data. The decision is based on a number of factors like the credit history of the applicant, frequency of transactions, kind of business, etc. The algorithm not only approves or disapproves the loan application, but it also determines the interest rate, time period and credit limit. Hence, AI improves the risk assessment for the banks and provides efficient decisions. This would further help in reducing the losses faced by banks due to fraudulent transactions and delinquent customers. Machine learning is also used to prevent frauds in online credit or debit card transactions [4]. Financial service providers like MasterCard suffer more losses due to false declines than fraud [5]. So, MasterCard has employed AI algorithms to learn the purchasing habits of cardholders in order to minimize false declines and to maximize the probability of identifying a fraudulent transaction.

Online shopping portals like Amazon and Flipkart have gained popularity in the last few years. These portals use efficient search engines to return the most relevant products to the user based on keywords. They also maintain the history of each buyer and recommend products based on their previous purchases. These portals also manage the demographic data of the customers, so that they can recommend products which were ordered by other customers with similar demographic data [6]. Amazon uses ANN to generate recommendations for its customers.

Nowadays, every smartphone is equipped with a feature that converts voice to text. About a decade ago, it was beyond imagination to accurately convert voice to text by the most advanced systems of that time. But ML has made it possible today, even for hand-held devices like smartphones, tablets, etc. Recently, Hinton et al. [7] proposed the use of artificial neural networks for speech recognition. Dahl et al. [8] proposed context-dependent pre-trained deep neural networks for large-vocabulary speech recognition. We start our day by saying "OK Google, open Maps", to find the best route from home to office. As soon as we start speaking, the ML algorithm instantly converts it into text and takes action. To connect a call, we only say "OK Google, call Mom" and it behaves like an obedient assistant. For Google's voice search, ANNs are used. At times, these speech recognition systems can transcribe more accurately than human beings. So, spoken instructions to smart personal assistants have now become the new interface. Siri [9] by Apple and Google Now are two personal assistants that can set reminders, call contacts from the phonebook, perform searches and manage appointments. Amazon took speech recognition one step further, by introducing complementary hardware called Alexa [10]. Alexa is a personal assistant that recognizes voice commands to carry out a number of tasks. It is powered by AI algorithms. Alexa can answer questions asked in natural language, set reminders, order food, play music and perform a number of other activities.

8.2.1 The Scope of Merger

All the applications of AI, discussed in the previous section, require huge amounts of data. A lot of research is being done recently to improve the response time of algorithms. The need to efficiently handle big data is the real motivation behind the merger of blockchain with AI. Data analytics involve the processing of enormous data and drawing correlations between patterns in the dataset. Blockchain is a distributed ledger technology (DLT) that stores data securely and transparently [11, 12]. In contrast to the centralized systems of operations which have been mostly used to date, blockchain is based on a decentralized system. Through the use of decentralized database architecture, the procurement, authentication and maintenance of various operations are dependent upon the agreement of several parties instead of a single central authority. Blockchain not only makes operations transparent, but safer and faster also. Blockchain became popular with the emergence of cryptocurrency Bitcoin [13], where transactions are managed using a distributed ledger on a peer-to-peer network that has anonymous, open and public access [14]. Blockchain is the underlying technology which maintains the Bitcoin transaction ledger. The blockchain transactions are stored in the public ledger and are verified by the consensus of a majority of participants in the system [15]. The information can never be erased in a blockchain as it maintains a verifiable record of every transaction. The digital currency Bitcoin itself is highly controversial, but the underlying blockchain technology has worked flawlessly and found a wide range of applications in both the financial and non-financial worlds. Figure 8.2 shows these two distinct technologies. Recently Salah et al. summarized upcoming blockchain applications and protocols while targeting artificial intelligence [16]. Dinh and Thal [17] compared the features of blockchain and AI. They suggested that the

Figure 8.2 Artificial intelligence and blockchain.

integration of these two technologies would revolutionize the new digital generation. Chen et al. recently proposed an AI convolution neural network-based nodes selection in blockchain networks [18].

8.3 Blockchain with Exonum Framework

Exonum [19] is an open-source blockchain framework that permits wide read access of blockchain data to the applications. It employs service-oriented architecture (SOA) [20] and consists of three parts: Services, clients and middleware. Here, services take care of the business logic of blockchain applications and are intended to implement logically complete and minimum functionality for solving a particular business task. Clients originate most transactions and read requests in a blockchain. Middleware is responsible for managing transactions, interoperating services among clients, access control, generating responses for client's read requests, etc. Exonum has the advantage that it is easier for clients and auditors to audit the system in real time. Due to the adoption of SOA, the application could easily reuse, add or configure the services developed for other Exonum applications. Exonum provides a significantly higher throughput capacity (order of 1,000 transactions per second) as compared to permission-less blockchain, and can encode complex transaction logic. Exonum uses pessimistic security assumptions for validator node operation. The consensus algorithm employed in Exonum does not introduce single points of failure. Furthermore, the set of validator nodes is reconfigurable, allowing it to scale the security by adding new validators, rotating keys for validators, locking out compromised validators, etc. Figure 8.3 shows Exonum service design where each service and auditing instance has a local replica of blockchain storage to ensure authenticity of the data and balancing the load.

Although blockchain emerged with the birth of cryptocurrencies, now blockchain has been infused into almost every possible industry [15, 22] including financial, healthcare, music, Internet of Things (IoT), quantum computing, and many more. The two emerging technologies, i.e. blockchain and AI, seem poles apart and

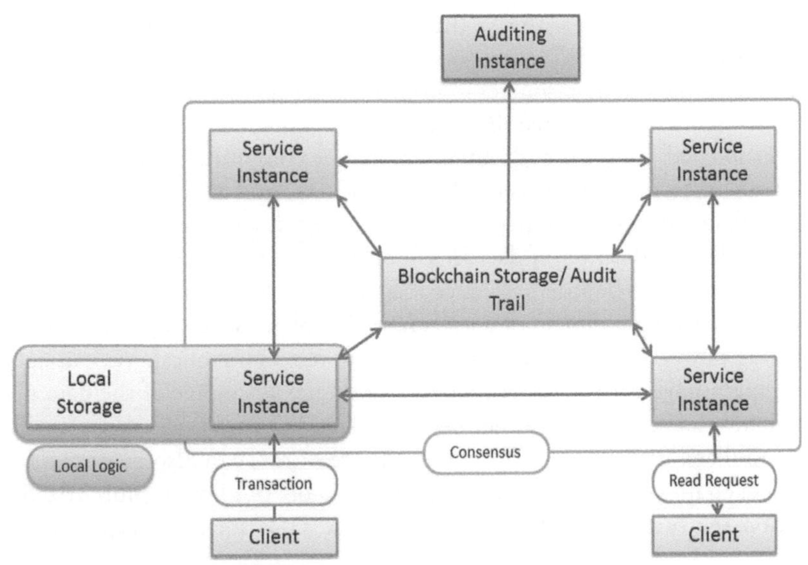

Figure 8.3 Exonum [21] service design.

they actually are. But the explosion of data in recent times has created a need for the fusion of these technologies. There are various startups like Endor and Blockchain Data Foundations that are actively working on the idea of the amalgamation of blockchain with AI. Neuromation [23] is a successful startup and proposes the use of distributed computing along with blockchain proof-of-work tokens to revolutionize AI model development. These startups have raised good amounts of funding from the investors, which itself proves that people are confident about the scope of blockchain and AI.

8.4 AI and Blockchain: Technologies That Are Poles Apart

The two technologies, artificial intelligence and blockchain, have contrasting features. So, before discussing the prospects of the merger, let us see the challenges in mixing them. If we look at AI and blockchain individually, their philosophies are different from each other. At the same time, it seems that their combination would be able to provide robust solutions.

8.5 Location of Control: Centralized or Decentralized

AI and blockchain have different thoughts about the location of control in the system. AI depends on very large, complete datasets with high configuration hardware to train the algorithms. The greater the size of the data, the better the performance of the AL/ML algorithm. Companies with huge data and enriched AI technology have more resources to improve and experiment with AI algorithms. This gives us a view that AI requires the centralization of data. It requires complete and consolidated data at one location for better computing results. In contrast, blockchain believes in the decentralization of data and its control. At the same time, the data and other resources are available to every user connected to the network. The decentralization of data in a blockchain involves execution cost due to network latency, and in order to reduce it, steps must be taken to speed up the data access. Blockchain guarantees that users have complete ownership of their data and computing power, which can be rented to other users on request. Various agencies and companies working with AI models could take advantage of the data present on the blockchain. They might not own the data, but can access them legally due to decentralized distribution. The computing resources can also be decentralized. Thus blockchain could manage the requirements of AI, allowing authorized users to develop and run AI algorithms on huge datasets provide by multiple users.

8.5.1 Visibility: Transparent or Black Box

AI and blockchain differ in their philosophies of keeping the data and transaction transparent. Blockchain is based on the principle of transparency. All the authorized users can view the ledger in a public blockchain. Even the anonymous data are available on the ledger. All the transactions are transparent and visible to other users also. Blockchain has another important principle, trust, which exists due to the use of cryptography. On the other hand, AI and ML algorithms are very difficult to understand. Most of the machine-learning algorithms require knowledge of advanced mathematical topics like optimization, linear algebra, etc., and are often considered a black box by most of their users. It is believed that in future blockchain could help in better understanding of AI-ML algorithms. Blockchain can provide public access to the data by which the AI models are trained and can analyze the potential weakness in the model. Since the performance of any AI model is based on the quality of the data, therefore a public audit of data can improve the correctness of data. This would further enhance the performance of the AI algorithm.

8.6 Blockchain-Facilitated AI

Artificial intelligence trains machines to behave like humans, for efficient decision making. The AI algorithms require extensive data for training and efficient decision making. Here, blockchain could play a significant role by storing and managing the data securely. Unlike traditional centralized database systems, blockchain creates a decentralized-distributed network of databases. This essentially means that the data are stored on a large network, where information is verified by multiple parties, and once data are entered, they cannot be erased from the blockchain. Any illegal mutation in data will be noticed by other computers of the blockchain, which will rectify the invalid data. Also, the data on the blockchain are secured through cryptography, which makes it incredibly difficult to decipher and modify. Any tempering with the data on the blockchain can be observed by the users due to cryptographic signatures associated with the data available on the network, across all the systems. This makes blockchain ideal for storing sensitive information. Therefore, blockchain is considered more secure than traditional database systems which are more prone to cyber-attacks. It appears that the requirements of AI can be supported by blockchain and we can expect a seamless communication between the two technologies. A few key characteristics of their interaction are discussed here.

8.6.1 Secure Data Sharing

The decentralized database of blockchain emphasizes secure data sharing between multiple parties since AI relies on large datasets, most of which will be shared. The performance of AI algorithms improves with the size of data. So, the availability of more data for analysis means better predictions by the algorithms and more reliability. The data supplied as input to AI algorithms should be valid. This requires a high level of data security, achieved by reducing the probability of any catastrophic occurrence. Blockchain also deals with sensitive data and has protocols that support data security [24]. The performance of AI algorithms could improve with the availability of more real-time data from multiple sources. Companies like Facebook, Google, Amazon, Flipkart, etc., have huge amounts of data, increasing every second, as shown in Figure 8.4. These data could be useful for a number of AI

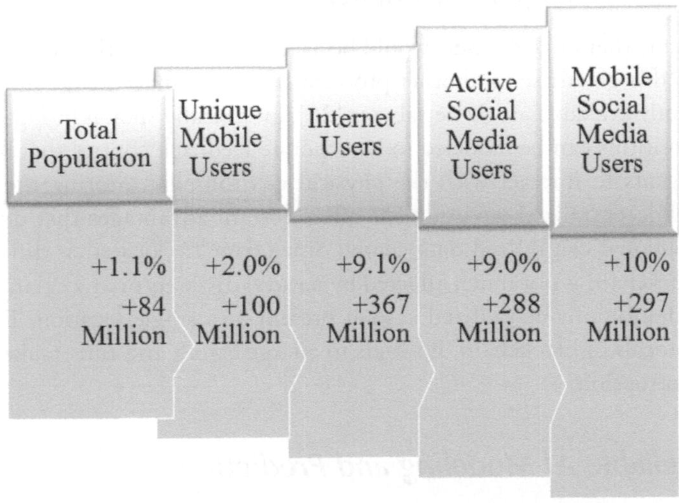

Figure 8.4 Annual digital growth (Jan 2018–Jan 2019): Change in statistics [25].

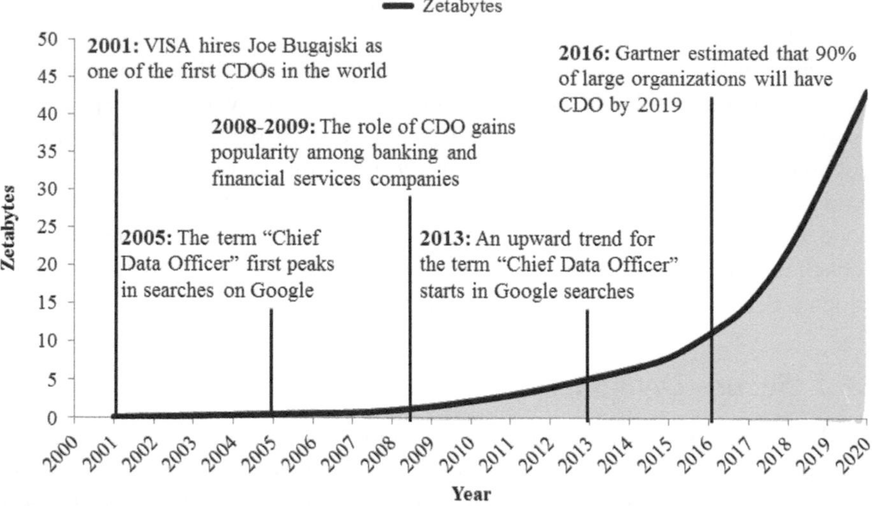
Amount of Data in the World
Zetabytes
2001: VISA hires Joe Bugajski as one of the first CDOs in the world
2005: The term "Chief Data Officer" first peaks in searches on Google
2008-2009: The role of CDO gains popularity among banking and financial services companies
2013: An upward trend for the term "Chief Data Officer" starts in Google searches
2016: Gartner estimated that 90% of large organizations will have CDO by 2019
Zetabytes
0
5
10
15
20
25
30
35
40
45
50
2000 2001 2002 2003 2004 2005 2006 2007 2008 2009 2010 2011 2012 2013 2014 2015 2016 2017 2018 2019 2020
Year

Figure 8.5 Amount of data in the world [26].

problems, but they are not freely accessible to public. This limitation of data accessibility can be resolved by blockchain, by the concept of peer-to-peer communication. Since blockchain is an open distributed registry, its data are available to every network user. Hence, the monopoly of companies over the data would be relaxed and other users could gain access to the data.

8.6.2 Data Management Challenges

Let us assume that huge datasets would be available to AI algorithms due to blockchain, but this gives rise to another problem of data management. The estimated data currently available are 32 zettabytes [26], graphically shown in Figure 8.5.

AI algorithms can be modeled to incorporate feedback control that helps independent agents to interact with the physical environment. Storing the data in a distributed-decentralized environment offers certain advantages that do not exist in a conventional centralized data center. Since data are located at different locations, a blockchain is not much affected by natural disasters or any calamity, which generally destroy any centralized system present at a single location. Due to the robust protocols of blockchain, it avoids to a large extent and thus makes data less prone to corruption.

8.6.3 Reliable AI Modeling and Predictions

The performance of any artificial intelligence depends heavily on the quality of input data stream supplied to it. It follows the garbage-in-garbage-out (GIGO) principle of computer systems. There are malicious users who intentionally modify the data for their individual or company's interest. Modified data would lead to the alteration of results also. Data may get corrupted unintentionally in instances like incorrect data captured by a malfunctioning sensor or a part of critical data may be lost due to hardware failure. Block can help in creating partitions of verified datasets, so that the AI models should use only the verified data. This will further identify any fault in the data chain and would lead to its correction. This would also help in reducing the efforts required in troubleshooting and locating the erroneous data in the segments. The data once captured through a transaction in a blockchain becomes immutable i.e. any intentional or unintentional alteration in the data will be reported by other users of the system. This makes the data traceable and verified.

8.6.4 Better Control over Data and Models

Control plays a major role in the success of any process. The alliance of blockchain and artificial intelligence provides better control over the data as well as the models. If a user creates data for his AI model, he can specify restrictions or permissions in the license for data. Licensing of data can be easily done in a blockchain. The permissions to read or access the data can be easily set in a blockchain. SingularityNET [27] is an open-source AI marketplace that collects smart contracts for a decentralized market of coordinated AI services. As per the SingularityNET team, blockchain technology provides transactional and bookkeeping advantages for managing network transactions. Blockchain allows users to add or upgrade an AI service for use by the network and get back network payment tokens in exchange. With this platform, the data owners can control the use of their data and will be benefited in monetary terms. Thus, blockchain, together with smart contracts [28], ensures that the data whether generated by financial transactions, applications or customer details, are valid, recorded in real time and are immutable. This ensures that correct and accurate data are available to be used by AI models and results in faster and secure transactions.

8.6.5 Real-Time Data

AI has transformed various domains like healthcare, finance, weather forecasting, etc., by providing real-time access to data. But these models suffer due to limited access to relevant data. Required data may not be available due to ownership restrictions. As a result, AI algorithms may end up working with low-quality data and the results may not be accurate. Blockchain could overcome this situation, as it provides access to a huge accumulation of data, which might be owned by distinct users and is immutable but accessible to all. Various applications designed using AI suffer from barriers like lack of authentication, unwanted intermediaries, risk of fraud, monopoly ownership on data, inaccuracy, etc. AI together with blockchain can reduce the ownership of a single entity, eliminate intermediaries and assure that the data is secure, accurate and authenticated by the stakeholders involved.

8.6.6 Transparency and Trust

Blockchain can add more transparency to AI by maintaining the details about any decision taken by the AI model. This information would be accessible in real time. The interested parties could look into the details available on blockchain, to find the fault in any wrong decision taken by the AI model. With more data availability, the issues can be resolved and algorithms can be improved. AI has revolutionized the modern industries, but it suffers due to lack of trust. Blockchain can be more reliable in building trust as it maintains a publicly accessible registry of the data and AI models, which are immutable and are digitally secured through cryptographic signatures [29]. The blockchain users are permitted to access the verified information and consensus models in real time. This would help in improving the trust of users on the system and would reduce the need for intermediaries.

This indicates that AI technology will be benefited by the robust blockchain. Now, let us understand how AI could improve the functioning of blockchain.

8.6.7 Better Management of Blockchain Using AI

Artificial intelligence could make the blockchain operations more efficient, by chaining the way they are managed. For instance, if a blockchain user transfers a Bitcoin to a peer on his network, the transaction might take days to get confirmed. This happens due to the decentralized nature of blockchain where the Bitcoin miners group the transactions into blocks. A sudden increase in volume of such transactions could further delay the transaction as the block-size is limited. If any transaction requires a block of larger size, it will be kept in queue for confirmation by the other miners. AI could reduce the complexity of this situation by improving the confirmation time of the transactions. Blockchain uses hashing algorithms [30] with a "brute force" approach that tries to find every possible combination of characters until the best match is found for the verification and confirmation of a transaction. The hashing algorithm could be replaced by an AI model which can be trained with correct data for efficient and quick confirmation of transactions. AI can improve the data mining process and can streamline the operations of a blockchain. This would further reduce the time and processing power required in mining the data in a blockchain.

Sharma et al. [31] explained the blockchain scaling problem where the size of blockchain is growing at 1 MB every 10 minutes and the existing data are around 85 MB. Blockchain has not incorporated any method to deal the data optimization and elimination. AI can help blockchain in this regard, by introducing a decentralized system for data optimization. The peer-to-peer transactions on a blockchain cost millions of dollars. The cost is high as each node executes the same task on its copy of data and tries to produce the solution before other nodes. AI can reduce the cost involved by predicting the node that will be the first one to deliver the solution. It can inform other nodes to shut down their operations, so that only the node which is going to give the earliest correct solution will complete the execution. This will ultimately reduce the cost of operations of the entire blockchain and can improve the efficiency of the system. Blockchain is known to maintain data with high security. AI can assimilate features like multi-dimensional data representation, image labeling, natural language processing, sound recognition, etc., into blockchain. This would allow miners to convert a large-scale system into several smaller subsystems that can optimize data transactions in a more secure environment.

In future, blockchain would store the data and interested users could buy them directly from the data owners. In this case, AI can provide solutions for keeping track of data usage, granting permissions and effective data management. AI models would behave as doorways through where the blockchain data would flow.

8.6.8 Prospective Applications of Blockchain and AI

It is already discussed that the philosophies of AI and blockchain are different. AI believes in centralized, fast and complex operations on huge datasets whereas blockchain provides decentralized, transparent but slow access to data and computing resources. Although blockchain is a robust technology, it suffers due to slow speed. Researchers are working on scaling solutions for blockchain to improve the speed of operations. If we integrate these contrary but complementary technologies, we could come up with interesting applications.

8.7 AI with Distributed Computing

The integration of AI and blockchain would lead us to the idea of using mining networks to provide computing power to the AI algorithms for training. These algorithms depend heavily on processor utilization for their extensive training where thousands of training sessions are conducted in order to teach the algorithms to make efficient decisions. The training time for ML algorithms is usually very high. If the system could provide more computing power to the algorithms then the algorithm could be trained in a faster way. So, the researchers and companies are working on the idea of boosting the AI technology with numerous GPUs, which are already used for mining cryptocurrencies. The GPU networks used for blockchain can also be used to provide more power to the ML algorithms. Another advantage of using GPU mining networks is that any user can access a distant supercomputer for executing hidden AI algorithm.

8.7.1 Data Privacy

Blockchain stores anonymous data, although it has ways to determine the source of data. The anonymous data provide great opportunities for research. Rather than centralized and restricted control of the companies over the data, blockchain provides open access to anonymous data, which could be used for analysis, model training and for making predictions. If the algorithms are trained with decentralized and anonymous data, they would be able to generate models with less bias. A company in the US might train the algorithms with the data collected from its population. With the use of blockchain, the data would not be restricted to a local region only. Hence, the algorithms would be trained on data that represent the global population.

8.7.2 Marketplace for Algorithms and Data

Open access to data and computing would allow AI researchers to work as a community. Many open-source packages are available for ML models like Scikit-learn, TensorFlow, etc. Blockchain would create an environment for a marketplace of data and algorithms. Data scientists and data owners would get paid for their algorithms and data respectively. Companies and individual users can access algorithms and data. An open environment can be created where AI algorithm development would not be proprietary, rather there would be marketplace where any interested user could participate and contribute. This would provide users with better control over the data. Today, companies like Google collect data from their customers without their consent like our search history and interaction with Google products like Maps, YouTube, etc. Those free data are used by Google to create information and to make money. A blockchain would give back the control of data to their owner. The data could be shared in exchange for certain services or could be sold to companies. This marketplace would allow anyone to generate revenue by being online and creating personal data which are worth selling.

8.7.3 Decentralized-Autonomous Organizations

Mohanty et al. [32] talk about the idea of decentralized autonomous organizations (DAOs), which is an application of AI-blockchain. DAOs hard code their rules, in the form of smart contracts. These organizations perform actions but follow the rules of a smart contract while taking any decision. By integrating AI technology with DAOs, there is a possibility that they can take independent decisions. The decisions would be based on market data and trends. Since DAOs exist on a decentralized network, there is a remote possibility that they will shut down. If they are developed well, they could reduce workload for knowledge workers, and would lead to economic equality with better management.

8.7.4 Use Case: Blockchain and AI in Healthcare

Health care providers, all over the world, are keeping track of patients' history through an electronic medical record-keeping system. These systems generate terabytes of medical data which are a gold mine of health information [21]. The data about patients including clinical tests, molecular diagnostics, MRI, electrocardiograms, etc., have significant value, depending on their quality and significance for certain conditions. Doctors study and analyze these data to assess a patient's condition at different time periods. Other than textual reports, data like images, videos, voice, etc., have considerable value in predicting a patient's medical condition. Recently, researchers have studied the use of voice and speech recognition for the diagnosis of Parkinson's disease and assessing its severity [33, 34].

Although medical data are available, the synergetic effect of the combinations of different data is often poorly understood. Here, we foresee the creation of specialized jobs for data economists who could understand the patterns in data from multiple perspectives. AI researchers are actively working for the creation of accurate prediction models that are based on input features like demographic data, laboratory tests and other diagnostic reports. The significance of a data value is based on the application. For example, an insurance company would be interested in a recent picture of a person, as it is a better indicator of his health, age and general status. Also, the combination of different data types will be more valuable than the individual data values.

A major problem faced by medical researchers is the availability and exchange of data. It is also important to maintain high standards of data security, due to the sensitive nature of the data. Blockchain could possibly reduce the occurrence of data breach or theft by involving multiple parties for data decryption, combined with the use of asymmetric cryptography. With a blockchain-backed system, the data can be easily uploaded by the patient (the owner) and he can give permissions in exchange for an agreed-upon price paid to him. But this comes with a challenge to devise a network currency that could overcome the problems due to having participants from multiple countries. Recently, Mamoshina et al. proposed a cryptocurrency called LifePound [21]. The data owner can generate this currency by putting his data on the blockchain and can be benefited by the incentive scheme. In this blockchain-based healthcare system, the blocks of transactions are processed and stored, along with management of keys for cryptography. The data are stored in encrypted form in the local storage. There are different categories of users. Some users are the owners of data and can upload and sell them in the marketplace. Others are consumers who are interested in buying the data. Some are data validators, and all these categories use LifePound as cryptocurrency. Figure 8.6 explains the architecture of the marketplace for medical data, backed by blockchain.

The clients of the medical data marketplace are of four types. The first category is of users who own the data. They upload their biomedical data and can sell it in exchange for cryptocurrency. They can keep their data private and can also provide

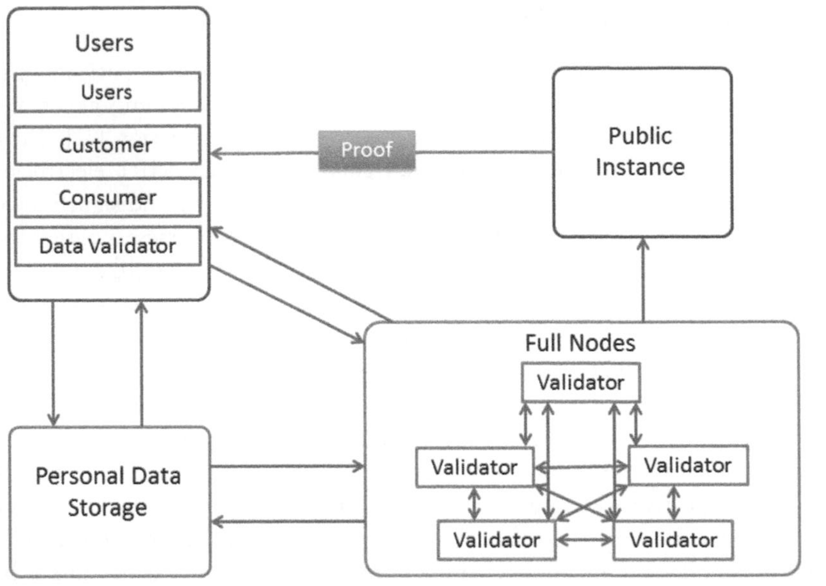

Figure 8.6 Medical data marketplace ecosystem [21].

the data anonymously. They will provide access to only those users who have paid for it. They could receive analysis of their health reports from data analysts working in the network. Another category of system users is the customers who are interested in buying the data for analysis and can provide reports to users about their health. They could get aggregate data collected from multiple users. Since it is an open system, it is important to validate the quality of data, which is done by data validators. They confirm the quality of the data and give assurance to customers. There could be other users who are using cryptocurrency.

Any interaction that occurs in a blockchain is registered in the form of a transaction. Blockchain maintains hashes which record a timestamp for a transaction and approve them. Timestamping is achieved through blockchain anchoring and the transactions are approved by digital signing and public key infrastructure. The storage of a medical data ecosystem could be an existing cloud storage like AWS that provides a platform for building applications [35]. This storage is not part of blockchain, but it is required to hold large medical files which are in the form of images or videos. To provide authentic access to the data, public key infrastructure is required which is established through blockchain. To keep the data secure, the data uploaded on the cloud, at the user side, are encrypted with a threshold encryption scheme [36]. The data can be kept at the user's local storage also which would provide actual privacy and control to the user over their data. There is a possibility of lending the data to customers instead of selling it. The full nodes of blockchain are the organizations that have complete access to data in the blockchain. They are of three types. Validators commit new blocks with transactions to the blockchain.

Another type is auditors who audit the marketplace. The third category is the key keepers who maintain key shares as per a certain threshold encryption scheme, required for decrypting the data.

Thus, blockchain allows the creation of a distributed and secure collection of personal data, where patients (data owners) have complete control over their data. They can also grant permissions to access their data. Thus medical data blockchain eventually builds a data marketplace where patients could get paid for their personal medical data. The data corpus thus accumulated will be used by data analysts, organizations for developing prediction models, pharmaceutical companies and by researchers. Currently, very few people go for comprehensive periodic medical tests that include laboratory tests, MRI, ECG, etc. If these data are combined with lifestyle data, demographic data and medical history, they would be of great value for researchers. There are number of companies that can pay for the data to train their AI algorithms. More data used for the training of an AI model means improved accuracy. The funds generated could be used further for research and development of sophisticated tools.

8.8 Conclusion

Blockchain and artificial intelligence are the two technologies that have the potential to revolutionize many industry sectors. We can foresee that the synergy of these two would have greater benefits. The fusion of AI and blockchain will bring growth in diverse domains and has many potential applications. The idea of their amalgamation is so powerful that it has recently attracted many researchers and organizations. Initially, AI and blockchain appear to be poles apart but at the same time, they complement each other. They have applications in fields like finance, medical, public service, security, banking, IoT, etc. The future of this integration is a decentralized system of operations, where machines could interact in a better way with better modeling of human activities.

References

1. Chang, Shiyu, Wei Han, Jiliang Tang, Guo-Jun Qi, Charu C. Aggarwal, and Thomas S. Huang. "Heterogeneous network embedding via deep architectures." In *Proceedings of the 21th ACM SIGKDD International Conference on Knowledge Discovery and Data Mining*, pp. 119–128. ACM, 2015.
2. Mitek Systems. https://www.miteksystems.com/mobile-verify.
3. Atiya, Amir F. "Bankruptcy prediction for credit risk using neural networks: A survey and new results." *IEEE Transactions on Neural Networks* 12, no. 4 (2001): 929–935.
4. Bolton, Richard J., and David J. Hand. "Statistical fraud detection: A review." *Statistical Science*, 17, no. 3 (2002): 235–249.
5. Tapscott, Alex, and Don Tapscott. "How blockchain is changing finance." *Harvard Business Review* 1, no. 9 (2017).

6. Bobadilla, Jesús, Fernando Ortega, Antonio Hernando, and Abraham Gutiérrez. "Recommender systems survey." *Knowledge-Based Systems* 46 (2013): 109–132.
7. Hinton, Geoffrey, Li Deng, Dong Yu, George Dahl, Abdel-rahman Mohamed, Navdeep Jaitly, Andrew Senior et al. "Deep neural networks for acoustic modeling in speech recognition." *IEEE Signal Processing Magazine* 29, no. 6 (2012): 82–97.
8. Dahl, George E., Dong Yu, Li Deng, and Alex Acero. "Context-dependent pre-trained deep neural networks for large-vocabulary speech recognition." *IEEE Transactions on Audio, Speech, and Language Processing* 20, no. 1 (2011): 30–42.
9. Assefi, Mehdi, Guangchi Liu, Mike P. Wittie, and Clemente Izurieta. "An experimental evaluation of apple siri and google speech recognition." *Proccedings of the* 2015 *ISCA SEDE*, 2015.
10. Ram, Ashwin, Rohit Prasad, Chandra Khatri, Anu Venkatesh, Raefer Gabriel, Qing Liu, Jeff Nunn et al. "Conversational ai: The science behind the alexa prize." *arXiv preprint arXiv:1801.03604* (2018).
11. Mills, David C., Kathy Wang, Brendan Malone, Anjana Ravi, Jeffrey Marquardt, Anton I. Badev, Timothy Brezinski et al. "Distributed ledger technology in payments, clearing, and settlement." (2016).
12. Iansiti, Marco, and Karim R. Lakhani. "The truth about blockchain." *Harvard Business Review* 95, no. 1 (2017): 118–127.
13. Radziwill, Nicole. "Blockchain revolution: How the technology behind Bitcoin is changing money, business, and the world." *The Quality Management Journal* 25, no. 1 (2018): 64–65.
14. Zheng, Zibin, Shaoan Xie, Hong-Ning Dai, and Huaimin Wang. "Blockchain challenges and opportunities: A survey." *Work Pap.*–2016, 2016.
15. Crosby, Michael, Pradan Pattanayak, Sanjeev Verma, and Vignesh Kalyanaraman. "Blockchain technology: Beyond bitcoin." *Applied Innovation* 2, no. 6–10 (2016): 71.
16. Salah, Khaled, M. Habib Ur Rehman, Nishara Nizamuddin, and Ala Al-Fuqaha. "Blockchain for AI: Review and open research challenges." *IEEE Access* 7 (2019): 10127–10149.
17. Dinh, Thang N., and My T. Thai. "Ai and blockchain: A disruptive integration." *Computer* 51, no. 9 (2018): 48–53.
18. Chen, Jianwen, Kai Duan, Rumin Zhang, Liaoyuan Zeng, and Wenyi Wang. "An AI based super nodes selection algorithm in blockchain networks." *arXiv preprint arXiv:1808.00216* (2018).
19. Yanovich, Yury, Ivan Ivashchenko, Alex Ostrovsky, Aleksandr Shevchenko, and Aleksei Sidorov. "Exonum: Byzantine fault tolerant protocol for blockchains."
20. Erl, T. *Service-Oriented Architecture: Concepts, Technology, and Design*. Pearson Education India, 2005.
21. Mamoshina, Polina, Lucy Ojomoko, Yury Yanovich, Alex Ostrovski, Alex Botezatu, Pavel Prikhodko, Eugene Izumchenko et al. "Converging blockchain and next-generation artificial intelligence technologies to decentralize and accelerate biomedical research and healthcare." *Oncotarget* 9, no. 5 (2018): 5665.
22. Mougayar, William. *The Business Blockchain: Promise, Practice, and Application of the Next Internet Technology*. John Wiley & Sons, 2016.
23. Neuromation. https://neuromation.io/neuromation_white_paper_eng.pdf.
24. Shrier, David, Weige Wu, and Alex Pentland. "Blockchain & infrastructure (identity, data security)." *Massachusetts Institute of Technology-Connection Science* 1, no. 3 (2016).

25. Annual Digital Growth, 2019. https://datareportal.com/reports/digital-2019-global-digital-overview.
26. Digital Data. https://liaison.opentext.com/blog/2017/06/06/ready-chief-data-officer-cdo/.
27. Goertzel, Ben, Simone Giacomelli, David Hanson, Cassio Pennachin, and Marco Argentieri. "SingularityNET: A decentralized, open market and inter-network for AIs." (2017).
28. Baars, D. S. "Towards self-sovereign identity using blockchain technology." Master's thesis, University of Twente, 2016.
29. Back, Adam, Matt Corallo, Luke Dashjr, Mark Friedenbach, Gregory Maxwell, Andrew Miller, Andrew Poelstra, Jorge Timón, and Pieter Wuille. "Enabling blockchain innovations with pegged sidechains." (2014): 72. http://www.opensciencereview.com/papers/123/enablingblockchain-innovations-with-pegged-sidechains.
30. Kraft, Daniel. "Difficulty control for blockchain-based consensus systems." *Peer-to-Peer Networking and Applications* 9, no. 2 (2016): 397–413.
31. Sharma, Pradip Kumar, Seo Yeon Moon, and Jong Hyuk Park. "Block-VN: A Distributed Blockchain Based Vehicular Network Architecture in Smart City." *JIPS* 13, no. 1 (2017): 184–195.
32. Mohanty, Soumendra, and Sachin Vyas. "Decentralized Autonomous Organizations= Blockchain+ AI+ IoT." In *How to Compete in the Age of Artificial Intelligence*, pp. 189–206. Apress, Berkeley, CA, 2018.
33. Wu, Y., P. Chen, Y. Yao, X. Ye, Y. Xiao, L. Liao, M. Wu, and J. Chen. "Dysphonic voice pattern analysis of patients in Parkinson's disease using minimum interclass probability risk feature selection and bagging ensemble learning methods." *Computational and Mathematical Methods in Medicine* 2017 (2017): 4201984.
34. Asgari, M., and Shafran, I. "Predicting severity of Parkinson's disease from speech." In 2010 *Annual International Conference of the IEEE Engineering in Medicine and Biology*, pp. 5201–5204. IEEE, 2010.
35. Gaetani, Edoardo, Leonardo Aniello, Roberto Baldoni, Federico Lombardi, Andrea Margheri, and Vladimiro Sassone. "Blockchain-based database to ensure data integrity in cloud computing environments." (2017).
36. Cramer, Ronald, Ivan Damgård, and Jesper B. Nielsen. "Multiparty computation from threshold homomorphic encryption." In *International Conference on the Theory and Applications of Cryptographic Techniques*, pp. 280–300. Springer, Berlin, Heidelberg, 2001.

Chapter 9

Identification of Blockchain-Enabled Opportunities and Their Business Values: Interoperability of Blockchain

N.S. Gowri Ganesh

Contents

9.1 Introduction

An application development evolved in various stages with respect to the evolution of the hardware and the software (operating systems). Thus it started with the applications that worked on independent computers whose benefits restricted to the users of that computer, distributed applications that delivered useful services to large number of clients with client/server architecture. Then with the introduction of web services the scenario changed from the tightly coupled application to that of the loosely coupled application. The loosely coupled application in the service-oriented architecture did not share the transaction and did not trust each other but relied on some centralized control.

The complete independence from centralized control with the introduction of the blockchain happened with the trust among the participants in the business networks. Also blockchain made a major impact by the introduction the concept of cryptocurrency like Bitcoin.

The evolution of the architecture in the field of application development and security has now transformed to the applications without agents or intermediaries (which increases the effective cost of a single transaction For example, transfer of money through bank involves commission by the agents) so that there can be peer to peer transaction in the decentralized network by establishing trust among unknown peers placed its position with the enablement of a new world of opportunities introduced with the concept of blockchain. Blockchain has given more integrity and independence to various applications by introducing the permissionless protocol and smart contract. The loosely coupled application has an improved form of automatic interaction with the implementation of the smart contract which has its basis in the traditional contract like SLA.

Blockchain is a decentralized computation and information-sharing platform that enables multiple authoritative domains to cooperate, coordinate and collaborate in rational decision-making processes. It is also described as an open distributed ledger that can note the transactions between two parties efficiently and in a verifiable and permanent way. Mining solves puzzles to add transaction records to the public ledger of past transactions. The ledger of past transactions is called the blockchain as it is a chain of blocks.

This ensures that no node has the power to sabotage the network and gain control. Present-day applications are developed in such a way that each participant in the business network keeps records, or a ledger, of all transactions between all the parties that the business interacts with. But blockchain allows any participant in the network to see the one system of record, or ledger.

The world is watching keenly the latest developments in the blockchain to identify the benefits that could be reaped with its implementation in their field of work. This chapter discusses the characteristics of blockchain in Section 9.3.1. The advantageous features of the blockchain instigate its integration in various fields of applications. The opportunities of blockchain are discussed as follows: Section 9.4.4 deals with the financial domain, Section 9.4.9 deliberates about its inclusion in healthcare systems; Sections 9.4.10 and 9.4.11 discuss its use in city management and in IoT devices. Section 9.6 is devoted to the software engineering practices specifically built with a blockchain approach.

9.2 How Does Blockchain Work?

Blockchain applications work in a decentralized manner. The nodes that are unknown to each other work together by sharing the data in the form of a public ledger. The data could be historical information of transactions like bank transactions that will be available to everyone for future computation. The public ledger ensures **consistency** which is nothing but that the local copies (Figure 9.1) of the data of all the nodes are kept identical and are always updated based on the global information.

Generally, in most of the applications, there are other modes of communications among the participating nodes such as centralized and distributed. In centralized architecture, a single server handles all the clients and therefore the failure of the server leads to the downtime of the application which is described as the single point of failure. The disadvantage due to single point of failure results in unavailability of the services (Baran 1964) (Table 9.1).

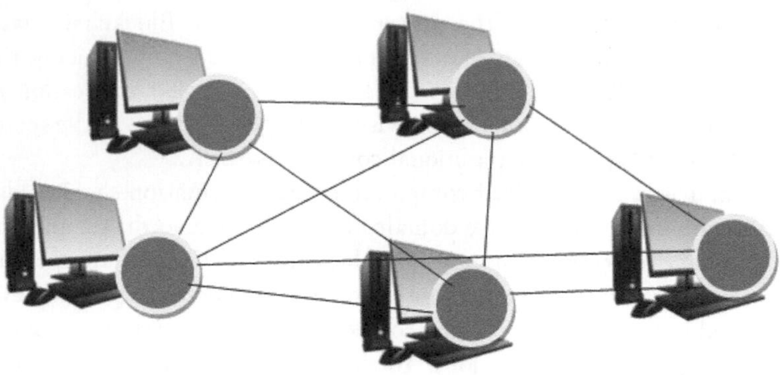

Figure 9.1 Blockchain maintaining local copies in participating node to ensure consistency.

Table 9.1 clearly explains how the decentralized nodes of Alice, Bob, Jane and Dave participate in the transactions and update the public ledger. As the transaction happens the information from the public ledger blocks transaction 4 when there is a shortage of money.

Formally **blockchain** (Iansiti and Lakhani 2017) can be defined as "*Blockchain is an open, distributed ledger that can record transactions between two parties efficiently and in a verifiable and permanent way*".

The following is a description of the flow of operations in a blockchain environment.

9.3 Flow of Operation

1) *Broadcast the transaction*: Each and every transaction is announced to the entire network for verification through broadcasting.
2) *Collection and verification of transaction*: The transactions are verified by the participants and will be added to the block according to the block size (1 MB for Bitcoin).
3) *Mining using consensus protocol*: The mining process is to solve the cryptographic puzzle. If a miner is able to solve the puzzle, this block is added to the blockchain. This is done using a consensus as proof of work (PoW) as in Bitcoin. There are consensus protocols used in blockchain like proof of stake (PoS), proof of burn (PoB) and proof of elapsed time (PoET).
4) *Updating the blockchain*: The nodes receive the block and can accept the block, provided that the computed PoW is correct and it contains valid transactions. Nodes add the block to their copy of the ledger. The hash of the last block will be used as the previous hash of the successive block. If solutions are obtained by two miners at the same time, the valid blockchain is the one which is

Table 9.1 Illustration of Transaction in Decentralized Public Ledger

Transaction	*Alice Public Ledger*	*Bob Public Ledger*	*Eve Public Ledger*	*Jane Public Ledger*
1.Initial	1. $50	1. $50	1. $50	1. $50
2.Alice send to BoB $25	1. $50 2. Alice→BoB:$25	1. $50 2. Alice→BoB:$25	1. $50 2. Alice→BoB:$25	1. $50 2. Alice→BoB:$25
3.BoB send to Eve $15	1. $50 2. Alice→BoB:$25 3. BoB→Eve:$15	1. $50 2. Alice→BoB:$25 3. BoB→Eve:$15	1. $50 2. Alice→BoB:$25 3. BoB→Eve:$15	1. $50 2. Alice→BoB:$25 3. BoB→Eve:$15
4. Alice send to Jane $30	1. $50 2. Alice→BoB:$25 3. BoB→Eve:$15 4. Less Money. So Trans. Blocked	1. $50 2. Alice→BoB:$25 3. BoB→Eve:$15 4. Less Money. So Trans. Blocked	1. $50 2. Alice→BoB:$25 3. BoB→Eve:$15 4. Less Money. So Trans. Blocked	1. $50 2. Alice→BoB:$25 3. BoB→Eve:$15 4. Less Money. So Trans. Blocked

longer. This demonstrates that the blockchain is tamper-proof and no transaction can be reversed. If the block of transactions or proof of work is invalid, the block is discarded, and the search for a valid block is continued by the nodes. Miners are allowed to earn rewards upon successful mining processes.

So blockchain works like the public ledger with the following characteristics.

9.3.1 Blockchain Characteristics

The blockchain can be generally characterized by the following concepts and varies for different implementations.

9.3.2 Asymmetric Key Cryptography

Blockchain maintains a digital wallet (equivalent to a bank account) secured with the user's private key and can be accessed with the signatures generated using that private key. This public key of the wallet serves as the address of the Bitcoin which is known to everyone and can be used for encryption during each transaction to protect users' privacy and anonymity. Transactions are digitally signed using the private key and the users keep it secret.

9.3.3 Block Containing Transactions

The blockchain is implemented on peer-to-peer architectures, and enables the exchange of information and sharing. The transfer information in the file is broadcasted to the entire network for validation from the source node. This set of transactions represents the current state of the blockchain.

9.3.4 Consensus in Blockchain Environment

Consensus ensures the correct set of operations in the presence of trustless individuals in a distributed environment. The properties of reliability and fault tolerance of a network are determined with consensus mechanisms. Each and every participating node agrees on a common content-updating protocol for their public ledger to maintain a consistent state. This is known as a consensus mechanism. Upon reaching consensus among the nodes the blocks are created and added to the existing ledger for later usage.

9.3.5 Mining for Block Acceptance

Miners solve the cryptographic puzzle for the acceptance of any block so that the block can be added to the shared ledger. The blockchain adds a nonce (number only used once) generated by the miners in the block which is a pseudorandom number that cannot be reused in replay attacks due to which the hashed or rehashed block meets the difficulty level of restrictions. The miners try to solve the puzzle to identify the nonce in the last block. The nodes accumulate the verified transactions in a block and use the computation power to find a value that makes the SHA-256 hash value of this block less than a dynamically varying target value. The header of the block includes the arbitrary nonce, the hash of the previous block, the Merkle root hash of the listed transactions, the timestamp and the block version. The hash of all the hashes of the transactions in a block is called the Merkle root. Every leaf node is the hash of the data block and the non-leaf node is the cryptographic hash of its child nodes. This is used in Bitcoin.

9.4 Classification of Blockchain Systems

Blockchain systems are classified as public, private or consortium.

9.4.1 Public Blockchain

A public blockchain is an open platform for participants from various backgrounds to join, perform transactions and mine. There is no constraint on any of these factors. These are also knowns as **permissionless blockchains**. Every user is given full authority to make transactions and perform auditing in the blockchain. The blockchain exhibits openness and transparency. It does not include any specific validator nodes. All users can collect transactions and participate in the mining process and earn mining rewards. The copy of the whole blockchain is synchronized with all of the participating nodes.

9.4.2 Private Blockchain

A private blockchain system is to aid the sharing and exchange of data among a group of individuals (in a single organization) or among multiple organizations privately. Here mining is controlled by single organization or selected group of individuals. This is known as a **permissioned blockchain**. This is because strangers cannot access it unless they are solicited. A set of rules decides the nodes' participation. This characterizes the network very much in the direction of centralization, while detracting from the blockchain features described by Nakamoto of decentralization and openness. But after nodes participate in the network, they contribute in running a decentralized network. The writes are restricted with the set of constraints.

9.4.3 Consortium Blockchain

A consortium blockchain is a partly private and permissioned blockchain, where a single organization is not responsible for consensus and block validation but rather a set of predetermined nodes. Consortium blockchains acquire the security features that are built into public blockchains. The consensus participants of a consortium blockchain are a group of pre-approved nodes on the network.

9.4.4 Blockchain Opportunities in the Financial Domain

Financial applications in the computing domain changed the dominant type of financial transactions from cash to digital money transactions with the help of banking automation. Further, the form of medium of monetary exchange is now introduced with the electronic wallets offered by many companies. In India the introduction of Unique Payment Interface (UPI) by the National Payment Corporation of India (NPCI) offered ease of money transaction. A relatively different approach to the electronic cash form is **cryptocurrency**. Bitcoin is a cryptocurrency which is the blockchain implementation form of digital assets that can be used as a medium of exchange with strong cryptography features to secure financial transactions.

9.4.5 Bitcoin

The term blockchain (Yuan and Wang 2018) was introduced when Bitcoin was first described in a posting in a cryptography mail group, about an article entitled "Bitcoin: A peer-to-peer electronic cash systems" (Nakamoto, n.d.) by a researcher with the pseudonym of "Nakamoto". Blockchain is defined as a kind of decentralized shared ledger that uses linearly timestamped, securely encrypted chains of blocks to contain verifiable and synchronized data across a peer-to-peer network.

9.4.6 Bitcoin Compared with Traditional Electronic Cash

Bitcoin can be compared with traditional electronic cash as listed below:

a) Bitcoin works in decentralized environment that applies distributed consensus algorithms in the participating nodes of peer-to-peer networks. The banks that use traditional electronic cashes require centralized service providers and are controlled by government agencies as per the mandate of the stipulated laws.
b) Bitcoin exhibits pseudo-anonymity. The identity of the Bitcoin user is not revealed as against the traditional electronic cashes where the users' identities are stored in the central server of the service provider.
c) Bitcoin is limited in terms of the currency issuance. The currency issuance in traditional electronic cashes is controlled by the forum designated by the government based on various factors like inflation rate, GDP, etc.
d) The software implementation of Bitcoin is open-source and is generally available to the users to check the algorithms. The business logic of the software

that issues electronic cash managed by the banks is closed sourced software executed in controlled secured environment and is not available to the users.

e) Bitcoin is a digitized form of zeroes and ones. However, Bitcoin can gain value by increasing users. The more users trust and use Bitcoin, the more value Bitcoin will have. In contrast, almost all traditional electronic cashes are endorsed by fiat money.

9.4.7 Bitcoin Ecosystem

Digitally signed and encrypted transactions are verified by peers. Cryptographic security ensures that participants can only view the information if they are authorized to do so. Data are grouped in a chain of blocks that are stored on all complete nodes of the blockchain network using a Merkle tree. The data structures of Merkle trees are encrypted, hashed and asymmetric time-marked. In particular, each node, after winning a consensus mechanism, will be allowed to place all data generated during that period (usually a regular interval, for example 10 minutes in the Bitcoin system) from the contest into a new timestamp unit, indicating the time of creation. If contradictory data are available, for example, Bitcoin double spending, only one agreed version is selected and added to the blocks.

The Bitcoin block consists of block headers and block parts.

a) The block header consists of (1) previous block hash, (2) Mining statistics used to construct the block (mining statistics – time stamp, nonce and difficulty), (3) Merkle tree root.
b) The block part consists of transactions.

Figure 9.2 shows the structure of the transactions in the form of a tree with a Merkle root. If a transaction is changed all the subsequent blocks need to be changed. The

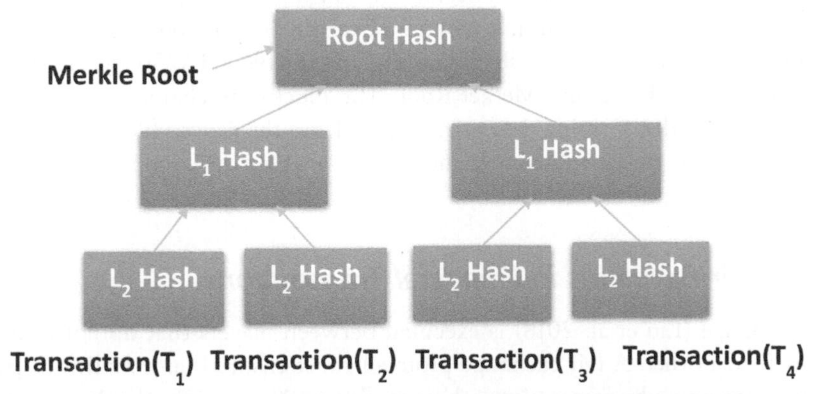

Figure 9.2 Hashed transaction with Merkle root.

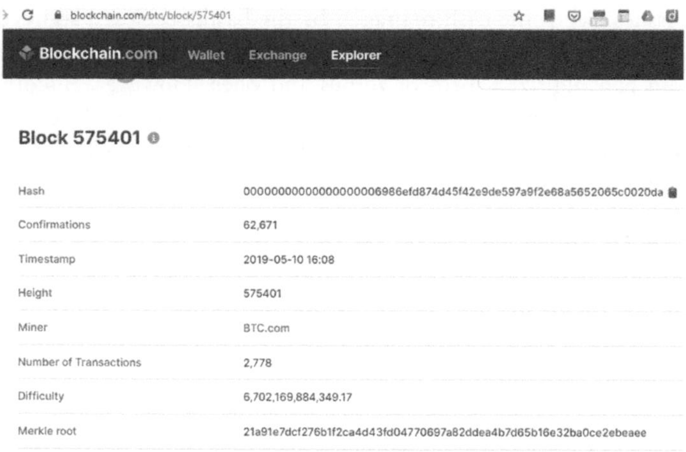

blockchain.com/btc/block/575401

Blockchain.com Wallet Exchange Explorer

Block 575401

Hash	0000000000000000000006986efd874d45f42e9de597a9f2e68a5652065c0020da
Confirmations	62,671
Timestamp	2019-05-10 16:08
Height	575401
Miner	BTC.com
Number of Transactions	2,778
Difficulty	6,702,169,884,349.17
Merkle root	21a91e7dcf276b1f2ca4d43fd04770697a82ddea4b7d65b16e32ba0ce2ebeaee

Figure 9.3 Snapshot of a blockchain summary for block 575401. (Source: https://www.blockchain.com/btc/block/575401.)

Hashes	
Hash	0000000000000000000006986efd874d45f42e9de597a9f2e68a5652065c0020da
Previous Block	0000000000000000000c5fd3157121e2a0f82d8468ecab1b91198fed2f1f5373
Next Block(s)	00000000000000000020bb6a82601d1b8e861b337518bf3e1f892670daa2f8f4
Merkle Root	21a91e7dcf276b1f2ca4d43fd04770697a82ddea4b7d65b16e32ba0ce2ebeaee

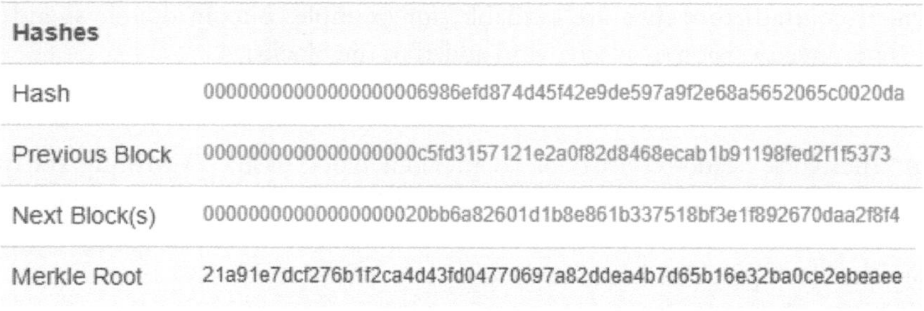

Figure 9.4 Snapshot of block header summary for the block. (Source: https://www.blockchain.com/btc/block/575401.)

difficulty of the mining algorithm decides the *toughness of tampering with the blockchain*. Figure 9.3 shows a summary of a blockchain transaction. Figure 9.4 shows the identifier of verified and hashed data (e.g. via double SHA256 algorithm), previous block, next block and Merkel Root. The blocks are chained one by one in chronological order, forming the entire history from the genesis block to the newly generated one. Figure 9.5 illustrates the transactions for the chosen block which is sourced from https://blockchain.com.

9.4.8 Fair Payments in Financial Transactions

Fair exchange (Liu et al. 2018) is executed between players that do not necessarily trust each other. A fair exchange protocol must ensure that a malicious player cannot gain any advantage over an honest player. We consider the "payment-for-receipt", where an entity, Alice, makes a digital payment to another entity, Bob, in order to get a receipt for the payment in the form of a digital signature. Our

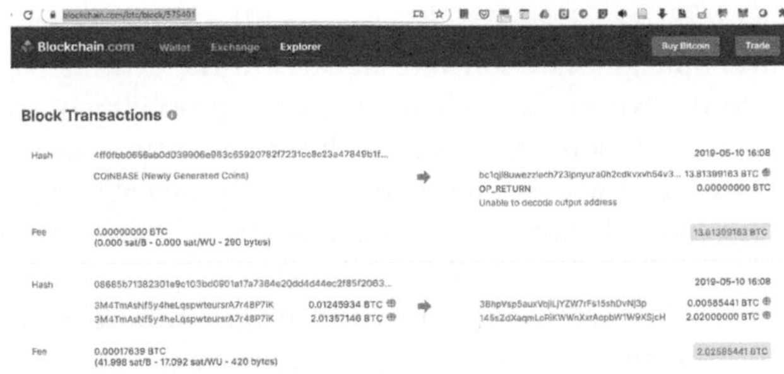

Figure 9.5 Transactions for the block 575401. (Source: https://www.blockchain.com/btc/block/575401.)

goal is to explore the solution space for integrating a fair exchange of payment for receipt into existing cryptocurrency payment schemes (hereafter referred to as fair payments for the sake of brevity). We assume the communication is weakly synchronous, under which messages are guaranteed to be delivered after a certain time bound.

9.4.8.1 Fair Exchange Using Timelocking

Allows the payer to pay in the time window if the payment has not been used. Alice produces a transaction that allows her to pay Bob a predetermined amount on the condition that Bob must issue a valid signature of a message within a certain period of time. The results of this transaction are input to one of the following two blockchain transactions:

1) A transaction signed by Bob that contains a valid signature on the requested message (that is, exchange is successful and fair).
2) A transaction signed by Alice and the time window has expired (that is, exchange fails and the money returns back to Alice).

9.4.8.2 Optimistic Fair Exchange Using Trusted Third Party (TTP)

The protocol is based on the existence of a trusted third party (TTP), but only in an optimistic way: TTP is only needed when a player tries to cheat. If Alice and Bob are honest and behave correctly, TTP generally does not need to be included.

9.4.8.3 Attacks in the Chain of Blocks

A fork (new branch of sequence of blocks) in blockchain is to impart change in the previous version or divergence from the existing protocol of the blockchain. Two types of forks can happen: 1. Software fork and hardware fork. Hardware fork is the one decided by a group of members who are interested to have an upgrade in the existing protocol (like change in the consensus algorithms, block size, etc.)

will come up with the new version of blockchain. The hardware fork is not backward compatible. Soft forks are software updates to the existing blockchain and are compatible to the existing blockchain. There is a possibility of malicious forking (aiming to create a fork in blockchain with the malicious need) and consensus mechanisms looks at destroying the other branch of a protocol fork merged mining which is used as a form of attack against a parent chain in the context of a hostile protocol fork.

The main idea of a pitchfork attack is to use merged mining as a form of attack on another branch in permissionless PoW cryptocurrency, which is the result of a controversial change in consensus rules. The pitchfork must reduce the usefulness of the affected branch so that the miner leaves the offending branch and switches to the branch of the fork which performs merged mining and follows the new consensus rules.

9.4.9 Blockchain Opportunities in Healthcare Systems

Blockchain technology (Casino, Dasaklis, and Patsakis 2019) has a major application in healthcare with several applications in various fields such as health management, health record management, automatic health claims, online patient access, use of patient medical data without loss of data privacy, drug counterfeiting, clinical testing and precision medicine.

The electronic healthcare patient (EHR) management record tends to be the area with the greatest growth potential. The uses of a blockchain-based system for EHRs are: There is no centralized server that can damage or interfere with hackers which results into single point of failure. Data are updated and always available, while data from various sources are collected in one repository. The health records are stored in a distributed way.

A number of projects focus on specific data modalities such as genomics and imaging. In particular, genomics met with great interest among entrepreneurs and businesses, perhaps because of the popularity of genome sequencing, the importance of genomic data and the great potential for monetization. Private genome companies such as 23andMe and AncestryDNA monetize genetic data by selling access to third parties such as labs and biotech companies. Some startup companies like Encrypgen, Nebula Genomics, LunaDNA and others are developing platforms or networks for genomics with data-exchange of genome using blockchain. With a blockchain-based platform, they claim to reduce the cost of genome sequencing, controlling patient data and sharing the value captured from the monetization of the data with patients.

Many proposed solutions are based on IP management blocks which can also be applied to innovation in drug development. An example of this is an attempt to use electronic lab notebooks from Labii using blockchain. Bernstein offers blockchain-based digital trail management with a time stamp to protect IP priorities. This function may be useful in collaborative pharmaceutical research. Solutions from iPlexus use the blockchain to make all unpublished and published data from drug development studies available.

Blockchain also has use cases in managing clinical trials for pharmaceutical research. The IEEE Standards Association held a forum for blockchain for clinical trials, exploring the use of the blockchain for patient recruitment innovations, to ensure data integrity and to make rapid progress in drug development. Scrybe, the blockchain project presented in the forum, provides an effective and reliable mechanism to accelerate clinical and research trials.

Parallel Health Systems (PHS) (Wang et al. 2018) uses artificial health systems to model and display patient status, diagnosis and care, then uses computer experiments to analyze and evaluate various treatment regimens, and carry out parallel implementations to support real-time decisions and optimization – both in actual and artificial conditions – in healthcare. The emerging blockchain technology is being used in the construction of PHS. In particular, consortium blockchain consisting of patients, hospitals, health offices, the medical community and medical researchers will be deployed, and smart and block-driven contracts will enable the exchange of electronic medical records (EHR).

MedRec and Patientory (Kuo, Zavaleta Rojas and Ohno-Machado 2019) offer the use of Ethereum-based blockchain platforms to exchange patient-driven health information. The same is adopted in clinical applications such as clinical data sharing and automatic remote patient monitoring. The clinical data exchange framework for oncological care patients recommends using the Hyperledger platform. Hyperledger is also involved in developing a framework for the implementation of Institution Review Board regulations. In addition, Hyperledger is available for mobile health applications and for medical data storage or access applications.

There are also some health-related blockchain applications that do not explicitly reveal their main platforms, such as Luna DNA, a database planned for genomic and medical research. The challenge of choosing a platform for health blocks consumes a lot of energy. Other problems include blockchain openness (e.g. public or private), the ability to modify and distribute code (such as licenses) and the need for certain hardware (such as SGX processors).

The Patient Centric Agent (PCA) (Uddin et al. 2018) includes lightweight communication protocols to enforce data security in various segments of continuous real-time patient-monitoring architecture. Architecture involves entering data into a private blockchain to facilitate the exchange of information among health professionals and their integration into electronic health records, while maintaining privacy. The blockchain was adapted for remote patient monitoring (RPM), requiring modifications that require PCA to select a miner to reduce the number of calculations. This allows PCA to manage multiple blockchains for the same patient and change the previous tree block to minimize energy consumption and to account for payment of safe transactions.

The center of the security mechanism is a key management system (Zhao et al. 2018), and an appropriate key management plan must be designed before the blockchains can be used in the healthcare system. Depending on the characteristics of the health blockchain, the authors here use a body sensing network to design lightweight backups and efficient recovery schemes for keys of blockchain. The authors' analysis shows that the system provides a high level of security and efficiency and can be used to protect messages that includes private information in health blocks.

Preventing prescription abuse (Engelhardt 2017): Prescription drug abuse involves well-defined challenges where blockchain technology can be applied. In one example, Nuco tried to use the three general approaches used to commit prescription fraud: Changing the number of recipe changes, copying/duplicating prescriptions, for example, "Doctor shopping", where scammers visit many doctors to collect as many original prescriptions as possible.

Nuco's blockchain-based solution for prescription abuse problems works as follows: When a prescription is given by the medical practitioner, it attaches a machine-readable code that functions as a unique identifier. This unique identifier is then associated with a block of information containing the name of the drug, the number, the patient's anonymous identity and the time stamp. If the prescription has been filled in by a pharmacist, the icon will be scanned, attempts to complete the prescription will be recorded and compared to the block, and the pharmacist will be notified immediately that the recipe meets the requirements of the padding and will provide information to verify the correctness.

The Nuco solution integrates existing usage patterns and uses existing technology (for example, pharmacists only need a smartphone or similar device to read unique identifiers), ensuring interoperability with existing protocols. Interoperability will be an important solution because the new blockchain has interfaces to existing projects, as well as new information storage technologies.

HealthChainRx and Scalamed are also working on blockchain-based anti-prescription fraud solutions. Healthcoin, an initiative that first developed a blockchain-based solution that allows people to work together to improve diabetes symptoms, has since expanded its vision to build a global electronic health record system.

HealthCombix tries to work with PointNurse by introducing nursing-mediated layers to ensure that data found in immutable block records are accurate enough to be transmitted correctly to patients. Patients understand how to prepare access to their medical records, updated and controlled.

Dentacoin is an initiative designed to use blockchain technology to connect dentists, patients and suppliers (producers and laboratories) throughout the world. Dentacoin provides the confidence and decentralization inherent in blockchains to develop economies of scale between participating parties without the need for additional intermediaries to manage interactions between each individual piece of the network.

Patientory (Katuwal et al. 2018) is one of the first blockchain-based healthcare startups to go for initial coin offerings (ICO) for funding. It developed HIE, supported by its own blockchain. HealthSuite Insights by Philips Healthcare examines the process of data exchange that is verifiable (ensures about the correctness of the data), a product that allows sharing of secure and traceable data between members of the hospital and university networks. All exchanges of the data inside the network are stored in a blockchain along with the identity of the person whose data is exchanged to create an audit trail from the exchange of data.

Medshare (Uddin et al. 2018) allows the blockchain-based data exchange of electronic health records among untrusted parties to be relied upon by entering source data, and auditing and tracking medical data. With smart contracts and access control systems, they claim that their systems can effectively track the behavior of data and decide to revoke access to the data based on incorrect data rules and insufficient permissions. Iryo is created for the global integrity of data repository in the OpenEHR format.

Healthcare Data Gateway (HDG) is a smartphone application that integrates traditional databases and blockchain distributed databases to manage patient health information. They propose a multi-party calculation (MPC) approach that allows third parties to access data but does not change data. Key management is based on a fuzzy repository on blockchain health architecture. The blockchain-based integrity framework architecture includes patient user sensor nodes, multiple implant nodes and body region sensor input nodes. The portal collects physiological data from supporting sensor nodes and sends aggregated data to designated hospitals that makes a block in the blockchain separately. The message generated by the gateway is considered as one block. Wearable sensor nodes produce keys before sending physiological data to the gateway node and encrypt data with keys generated by the patient's body signals. Neither the blockchain community nor health professionals can leak patient information. Patients can only recover keys from their physiological data to decrypt the data. However, this approach places a significant burden on limited power medical sensors, because these sensors must build key physiological patient data during decryption.

Patient Centric Agent (PCA) connects patients' body sensor network (BSN) to personalized blockchain networks. PCA decides which data should be included in the blockchain and which miners should be chosen. The blockchain is not only a distributed database for patients, but also an authentic platform that is checked by all nodes on the blockchain. The blockchain node can be provided by healthcare providers, other organizations or individuals.

A BSN (Zhao et al. 2018) consists of dozens of biosensor arrays located on or inside the human body. This node is equipped with various biosensors which can detect physiological signals such as blood pressure (systolic and diastolic), electrocardiogram, blood oxygen content (SpO_2), photoplethysmogram signals (PPG) and so on. In addition, they are also equipped with wireless network chips, and these chips not only assist biosensor nodes in forming BSN, but also help these nodes to send composite physiological signals to special relational nodes (commonly referred to as PDAs) that are responsible for combining and sending signals to remote medical centers, such as hospitals.

In September 2017, Axa, France's leading insurance group, offered parametric flight delays based on the Ethereum's fizzy™ platform, which uses smart contracts related to the global flight database. After a flight delay is detected, compensation begins immediately and safely, eliminating the need for additional documents. Such a program can be achieved using blockchain in health insurance, so medical records do not need to be examined and the process is very efficient.

9.4.10 Blockchain Opportunities in City Management

Digital governance (Shen and Pena-Mora 2018) contributes to the important agenda of sustainable development, such as reducing corruption, reducing administrative costs, ensuring document integrity and connecting donors and disadvantaged groups such as refugees and displaced people. To see how block technology, as digital technology, can have a significant impact on city management, it is helpful to start with the four ideal concepts of smart city management (Meijer and Bolívar 2016). They include: (1) government of a smart city, (2) smart decision-making, (3) smart administration and (4) smart urban collaboration.

There is also a blockchain system designed to change the two most important processes in government – one is the voting for the formation of the government and the other is government tax. The aim of an electronic voting system is to achieve anonymity, privacy and transparency. Anonymity ensures that voters' voices cannot be prosecuted. Privacy assures voters that their data is not misused, and transparency ensures that the election mechanism cannot be violated. The design of blockchain-based voting systems can be found in a number of studies. Some also produce prototypes. However, it is a question of all of these systems that voter authentication must be guaranteed at the personal level outside the blockchain.

In the tax field, the blockchain solution allows the tax authority to better control the tax system. A private blockchain can be managed by the tax administration to monitor VAT invoices and keep immutable information about taxable transactions to avoid tax revenue losses.

One project illustrating this vision concentrates on the area of urban policy making. The authors state that current urban codes such as policies, planning, regulations and standards are not up to meeting the urban sustainability challenges due to their top-down delivery and implementation methods. Blockchain-based mechanisms make it possible to truly deliver and execute urban codes bottom-up. In the case of policies and codes, citizens submit their urban needs to the blockchain, which will be prioritized by a blockchain consensus mechanism for the authorities to draft policies. These drafts will be ratified through the blockchain validation capabilities. Further transformation of these plans into physical forms (e.g. construction of infrastructure projects) can be approved via voting mechanisms on the blockchain as well. The plans and regulations can also be standardized for replicability and scalability purposes, using the same bottom-up approach for citizen participatory standardization.

The blockchain system has been proposed to support records of immutable educational processes. There are proposals that record creative work or ideas to gain a scientific reputation, keep a diary of student activities in various learning organizations and allow higher education institutions around the world to give credit to courses for students who have completed the exhibition of their ideas in the course. Training and other records can be included in the general system for managing personal records and used by companies and audit services.

Researchers also use blockchains to solve problems related to the academic community. Examples of applications cover the entire life cycle of research methodology, peer review and research publications for the protection of intellectual property. First, it is proposed at the experimental stage to block and release the data recording system and its results if necessary to avoid damage to the experimental integrity by negligence or intentional errors, e.g. audit trail from research data. This guide recommends the use of adaptive blockchain-based choreography for collaborative experiments, which can be reproduced silico experiments towards both robust accountable reproducible explained (RARE) research and findable accessible interoperable reusable (FAIR) results. Second, the paper phase introduces a blockchain-based platform that stores and measures author contributions based on changes made by the author. Third, the blockchain system in the peer review phase can also stimulate a timely and sustainable review process. As explained above, the system can give the examiners a cryptographic prize if the quality check is received by the editor. This valued currency can later be used to publish reviewers' papers in journals, creating an incentive mechanism. The fourth part uses the previous work of Semantic Web technology in the release phase to give the author the opportunity to work on an evolutionary version of scientific research that can be opened for reviews, conferences or journals. This allows decentralized publishing systems. These include blocking chains, smart contracts and MAS to coordinate the traceability of food in agricultural products. Implementing this new model will increase the supply chain of current agricultural products by adding blockchain. The current supply chain and supply chain architectures via blockchain models are described below, including the advantages provided by the new supply chain model.

(1) Current supply chain: This model starts with manufacturers and imports. Two supply chain members send their products and data to the next layer of the supply chain. The next shift is export, process or wholesale. This is the middle layer that processes the main products of the supply chain. Finally, the last layer accommodates retailers and catering service providers who sell products. The main disadvantage of this model is that data in each element of the supply chain are centralized and other elements cannot see the transaction. The main result of this error is that consumers do not have the opportunity to check the source of food to be purchased. In addition, there is no way to ensure that user data is reliable.

(2) Supply chain through blockchain: With the addition of blockchain to agriculture, the supply chain changes. Now, all supply chain members store all their transactions in one block. This allows greater security in transactions. In addition, this new model corrects the shortcomings of the current supply chain. Data are decentralized, and each member can read important data about the block operation. For example, the manufacturer can check processor product information and transport provider pickup details.

A traceability system (Lu and Xu 2017) enables product tracking by providing information (for example, origin, components or location) during production and distribution. Product suppliers and retailers usually require independent and government-certified tracking service providers to verify products throughout the supply chain. If all meet the requirements, the traceability service provider issues a test certificate confirming the quality and authenticity of the product. Tracking systems usually store information in conventional databases controlled by service providers. Centralized data storage is a potential single point of failure, and there is a risk of tampering.

This restructures the current traceability system to track service providers by replacing the central database with a blockchain. OriginChain provides transparent data tracking, increases data availability, and automates compliance testing. OriginChain is tested realistically based on user tracking information. Product suppliers and retailers request tracking services for various purposes. Suppliers want to receive certificates to show consumers their origin and the quality of their products and to comply with regulations. The dealer wants to check the origin and quality of the product.

OriginChain currently uses private blockchains that are geographically dispersed in tracking service providers, which have offices in three countries. The plan is to create a reliable tracking platform for other organizations, including government-certified laboratories, large suppliers and retailers with long-term plans. The parties signed a legal agreement covering tracking services. OriginChain produced smart contracts that represented legal agreements. The smart contract codifies the combination of services and other provisions specified in the agreement. A smart contract can automatically check and apply this condition. It will also check whether all information required by the regulations is available to allow an automatic review of compliance with regulations stipulated with the company.

Tracking service providers manage required information for traceability in place such as regulations, date of inspection, etc. and hash of certificates or photos. Because data memory is limited to the blockchain, originChain stores two types of data along with the blockchain as variables for smart contracts:

- The hash of traceability certificates or photos.
- The small amount of information about the traceability required by the regulation, such as the batch number, results of the traceability, origin and the date of inspection. The traceability certificates in the form of raw files and

photos (.pdf or .jpg) and the addresses of the smart contracts are off-chain. It is available in a centralized MySQL database hosted by originChain. Other partners may still have their own product information database (for suppliers or retailers) or other sample numbers (for laboratories). The laboratory periodically injects the results of test samples from the outside world into blockchain. Blockchain permission control can be off-chain or on-chain. However, the centralized off-chain rights management module can be the focal point of withdrawal, both operationally and from a management perspective. OriginChain stores control information, for example, information for having permission to join the blockchain network (having a copy of all historical transactions). On-chain permission management leverages the blockchain by influencing the nature of the decentralized blockchain, so that all participants have access to the blockchain. On OriginChain, the factory contract creates a smart contract. This reduces the complexity of creating special smart contracts. The employment contract contains code fragments that represent various tracking services. The creation of smart contracts requires permission from search service providers and suppliers or dealers. When a factory contract is invoked, it creates two types of smart contracts: Signing contracts and service contracts. A registration contract is a legal agreement and addresses the service contract that codifies the legal contract. Service contracts can be renewed by changing the address specified in the registration agreement with the new version address. Possible updates include adding or removing services from legal agreements after signing an original legal agreement or selecting a testing laboratory based on availability. The registration agreement contains a list of addresses that are permitted to renew the registration agreement and the threshold for the minimum number of addresses needed for upgrading.

9.4.11 Blockchain Opportunities for Security Measures in IoT Devices

Security (Kolokotronis et al. 2019) and privacy are increasingly important factors in the launch of Internet of Things products and services. Recently, there have been attacks where Internet devices were used to conduct distributed DDoS attacks, spy on people and hijack communication links so that the attacker has complete control over anything that they can remotely access. DDoS attacks, cloud-based and mobile are among the most common attacks. The availability of botnets for leasing has caused a significant increase in DDoS attacks, and it is very possible that IoT will further facilitate the creation of these botnets. The latest example of a DDoS attack, in October 2016, associated with the Mirai botnet, affected millions of users and businesses and influenced popular service servers such as Twitter, Netflix and PayPal. This simple malware infected Internet devices with things that use default settings and credentials. In October 2016, US provider DNS Dyn experienced a cyber-attack. Dyn attacks come from "tens of millions of IP addresses" and at least some traffic comes from IoT devices, including webcams, baby monitors, home routers and digital video recorders. Malware called Mirai, which controls online devices and uses them to initiate for DDoS attacks. The process involves phishing emails that infect computers or home networks. Malicious software then spreads to other devices such as DVRs, printers, routers and cameras that are connected to the Internet and used by businesses, and most companies are forced to compromise on surveillance issues. Tools like Shodan and IoTSeeker can be easily used to detect vulnerable devices. This raises important questions about how widespread use of such vulnerabilities can be prevented, because Internet-based devices have very limited self-sufficiency.

The creation and management of vulnerability profiles, which may involve manufacturers, can ensure that users are dealt with seriously in matters of security and privacy. Blockchain must define a new fundamental approach to security that goes beyond each other.

The device itself must contain the following:

Identity security: Blocking identity theft, prohibiting the use of unfair public key certificates, countermeasures against "humans in the middle"

Data protection: Prevent data manipulation, development of access control mechanisms and blockchain abuse

Security communication: Domain name services, DDoS attacks, important information infrastructure protection

In particular, the approach to public safety through transparency has clear benefits for the Internet of Things.

To improve the security of CE devices that support IoT, consider the following phases throughout their life cycle.

9.4.11.1 Registration

When assembled, the product goes into a blockchain that links its cryptographic trace with blockchain entries.

9.4.11.2 Update

If changes are made, e.g. as a firmware update, new fingerprints are created by the peers and sent to the network, which insert fingerprints using the consensus algorithm in their local copy.

9.4.11.3 Inspection

A node can check device properties at any time by recreating fingerprints and comparing those values with entries (correct) in one block.

However, this can be considered as the hope that the device behaves well for certain purposes and does not pose a threat or danger to the other party. Both objective (e.g. vulnerability, integrity, etc.) and subjective measurements (for example, recommendations or reputation) contribute to this proposal in calculations to use blocks in the form of crypto blocks as alternative bits (called altcoins), which are accompanied by consensus protocols adapted to the application. Examples include distributed access management systems where users own and control their personal information, binary systems and certificate monitoring systems and crypto bodies that can be used by the device to show that it has contributed to DDoS attacks on certain purposes. The security of this proposal, if managed strictly, depends on assumptions about the security of the basic block data structure.

When the blockchain is integrated into IoT technology, IoT devices exchange data across distributed books and smart contracts. In this scenario, each device can be disconnected, generating around a device's footprint, because each device leaves a unique track. So, when a device is associated with someone, personal data are processed. This is in line with the European Data Protection Regulations (GDPR) (EU 2016/679), according to which a pseudonym may not be considered anonymization even though it reduces the risk of data from the data subject that causes threats. It is assumed that GDPR applies to most organizations, even if they are not within the European union but their data is processed within the region of European Union jurisdiction of law.

Another challenge that may be faced by the blockchain on legal compliance is the deletion of personal data from books if the user (if any) cancels the processing authorization. This is called the right to be forgotten in GDPR.

The partnership between IBM and Samsung has produced an autonomous platform, decentralized for peer-to-peer exchanges. In particular, Ethereum is used to coordinate devices by providing features such as registration, authentication and consensus revocation lists. Gladius recently proposed an approach to reduce DDoS attacks using a blockchain where pool nodes are dynamically formed (through the Intelligent Ethereum contract) to validate the requested link and block malicious activity. Additional security tools for IoT blockers such as Factom, Filament and Guardtime have been developed to focus on protecting system component integrity.

With the integrated IBM Watson IoT platform, users can add selected blocks of registry data from the Internet to private registry blocks that may be included in shared transactions. The platform translates data from devices connected to the format required by the API for one block contracts. The blockchain agreement does not need to know the specificity of device data. This platform filters events on the device and only sends the data needed to execute the contract (ibm .co/2rJWCPC).

Creating new business models eliminate the need for centralized cloud servers, For example, Filament, a blockchain-based solutions provider for IoT, has launched wireless sensors, called Taps, that allow communication with computers, phones, or tablets within 10 miles (bit.ly/2rsxZYf). Taps create low-power, autonomous mesh networks that enable companies to manage physical mining operations or water flows over agricultural fields. Taps don't rely on cloud services. Device identification and intercommunication is secured by a blockchain that holds the unique identity of each participating node.

Business is the main reason behind the creation of blockchain-oriented software. The demand for security in blockchain applications is, therefore, even more pressing, and thus that for specialized software engineering processes.

9.5 Interoperability among Blockchains

Heterogeneous blockchain systems cannot trust or communicate with each other. You cannot exchange values. However, moving assets among ledgers leads to convenience. Consumers are increasingly interested in exchanging information between blockchains. Linking activities with different chains makes sense. For example, an institution can request the arrival of funds in the blockchain to make appropriate transfers of funds to others. In fact, there are a number of connectors that facilitate payments between these ledgers, and there are major obstacles to introducing new connections.

Gideon has proposed a multichannel that is easily configured and can work with different blocks. In addition, connections between chain work can be made. Pegged side chains proposed in Blockstream allow battle transmissions and other registration resources between several blocks.

To reduce the barrier between the ledgers that are heterogeneous decentralized, there is an expanded blockchain architecture called interactive multiple blockchain architecture. Side chains complement the Bitcoin protocol, allowing trustless communication between Bitcoin and side chains. The pegged side chain can transfer Bitcoin assets and other ledgers between many blockchains. Users can easily access new cryptocurrency systems with assets that are on other systems

Cosmos is a new architecture of networks. This allows parallel blockchains to work together while maintaining their protective properties. Many independent block networks are called zones. These areas are powered by a highly efficient, consistent and safe consensus engine. The first area of the cosmos is the center of the network. This acts as the government of the entire system, which allows the network to adapt and modernize. In addition, hubs can be extended by connecting other zones. Zones provide future compatibility with the new blockchain because each blockchain system can connect to the Cosmos hub. It can also isolate everyone from the failure of other regions. Cosmos allows blockchain communication through protocols, such as some kind of UDP or virtual TCP. Coupons can be safely and quickly transferred from one zone to another without having to exchange liquidity between zones. To monitor the total number of tokens held by zones, all tokens go through the Cosmos Hub.

Polkadot is a collection of independent chains with integrated security and shared interchain transactability without trust. Applications provided for Polkadot must be parallel to the parachains. Each parachains is operated by another segment of the Polkadot network. Polkadot leaves a lot of complexity investigated at the middleware level. In addition, it outlines multi-chain protocols that can be scaled with the ability to be blockchain protocol that are compatible with heterogeneous block systems that interact by creating dynamic block networks called router blockchains. The router blockchain contains a number of router nodes. Chains join a blockchain network before the node becomes the routing node that is a member of the router block. All routing nodes with details about different circuits become blockchain systems that support router information. After the router information is updated, all router nodes match the latest routing table. In this way, the router blockchain system records validated addresses from each participating blockchain. When a transaction is generated between circuit A and circuit B, circuit A can be connected to circuit B, the data sent correspond to routing information written in the router blockchain.

9.6 Building Blockchain-Oriented Software

Blockchain-oriented software (BOS) (Porru et al. 2017) is defined as the software implemented with blockchain. Blockchain is a data structure that is marked by the following key elements:

- Data redundancy (each node has a copy of the blockchain)
- Verification of transaction requirements before validation
- Record transactions in blocks ordered sequentially, making them controlled by consensus algorithms
- Transactions based on public key encryption
- Possible languages for transaction scripts

Software architecture: For the development of special BOS design records, a macroarchitecture or meta-model can be defined. For this purpose, software engineers must establish criteria for selecting block performance that is most appropriate for assessing acceptance of side chain technology or ad-hoc block applications. For example, Ethereum5 has received key storage, which is a very simple database. Using a higher level of data representation, for example, graphical objects, can speed up many operations that should be expensive with key-value repositories.

Modeling language: Block-oriented systems may require special graphic presentation models. In particular, existing models can also be adjusted to BOS. UML diagrams can be changed or even rebuilt to reflect BOS features. For example, charts such as Case Use, Activity Diagrams and Status Charts cannot effectively represent the BOS environment.

Metrics: BOSE can utilize the introduction of certain metrics. For this purpose, it will be useful to refer to the target/metric/metric (GQM) method, which was originally developed to make measurement activities, but can also be used to control analysis and improve the software process.

IBM recently stated the need for ongoing testing to ensure the quality of block software. Testing must be based on the type of application, which in the case of BOS is a critical safety system. In particular, the applications must be tested for BOS. These testing packages must contain:

- Smart contract testing (SCT), in particular special testing to verify that smart contracts (i) satisfy the principals.

Specifications, (ii) comply with jurisdictional law and (iii) do not contain unfair contractual provisions.

Testing of blockchain transactions (BTT), e.g. double spending test and condition integrity (e.g. UTXO4).

Making software for smart contract languages: Implementation of the smart contract development environment (SCDE) – block-oriented IDE transfer can be key to building and disseminating BOS knowledge. Such an environment can facilitate intelligent contracting in special languages (for example, Solidity, language for writing contracts in Ethereum).

Blockchain-oriented software engineering (BOSE) (Wessling and Gruhn 2018) is a new research field for decentralized application development ("DApps" for short) based on blockchain technology. Currently, Ethereum blockchain is the most popular platform to build DApps. The business logic is represented by one or more executable code contracts (abbreviated as "EDCC", the term used to describe smart contracts) located on the blockchain network. This involves designing existing DApps, identifying possible architectural models and comparing their pros and cons with the basic architectural model of DApp, where users interact directly with EDCC by generating and sending transactions. There are three ways:

- Self-generated transactions: Users can (1) send transactions directly to blockchains, (2) use web interfaces such as MyEtherWallet or (3) use an integrated wallet browser like Chrome with MetaMask [9], or a crypto-browser like a cipher or status (each variant can be executed with a private blockchain running on a user's device or public node managed by a third party such as Infura or Etherscan).
- Self-confirmation transaction: Interaction with DApp is mainly done with Cryptobrowser or MetaMask. Transactions are not generated by the user but are triggered by the DApp website, presented to the user for further review and then manually sent to the blockchain node. In this way, this model offers a compromise between convenience and trust needed on the DApp website and transaction details.

The DApp provider provides websites that can interact with users without requiring Cryptobrowser or the MetaMask plugin. The website communicates with the DApp logic backend through REST calls and summarizes all blockchain-specific actions. This means that the backend is responsible for interacting with the blockchain and sending transactions to users who cannot validate it. For this reason, Pattern C offers maximum comfort, but places great trust in the DApp provider, which processes user input data and manages the private key.

References

Baran, Paul. 1964. "On Distributed Communications." Product Page. 1964. https://www.rand.org/pubs/research_memoranda/RM3420.html.

Casino, Fran, Thomas K. Dasaklis, and Constantinos Patsakis. 2019. "A Systematic Literature Review of Blockchain-Based Applications: Current Status, Classification and Open Issues." *Telematics and Informatics* 36 (March): 55–81. doi:10.1016/j.tele.2018.11.006.

Engelhardt, Mark. 2017. "Hitching Healthcare to the Chain: An Introduction to Blockchain Technology in the Healthcare Sector." *Technology Innovation Management Review* 7 (10): 22–34. doi:10.22215/timreview/1111.

Iansiti, Marco, and Karim R. Lakhani. 2017. "The Truth about Blockchain." *Harvard Business Review*, January 1, 2017. https://hbr.org/2017/01/the-truth-about-blockchain.

Katuwal, Gajendra J., Sandip Pandey, Mark Hennessey, and Bishal Lamichhane. 2018. "Applications of Blockchain in Healthcare: Current Landscape & Challenges." *ArXiv:1812.02776 [Cs]*, December. http://arxiv.org/abs/1812.02776.

Kuo, Tsung-Ting, Hugo Zavaleta Rojas, and Lucila Ohno-Machado. 2019. "Comparison of Blockchain Platforms: A Systematic Review and Healthcare Examples." Journal of the American Medical Informatics Association 26 (5): 462–78. doi:10.1093/jamia/ocy185.

Liu, J., W. Li, G. O. Karame, and N. Asokan. 2018. "Toward Fairness of Cryptocurrency Payments." *IEEE Security Privacy* 16 (3): 81–89. doi:10.1109/MSP.2018.2701163.

Lu, Q., and X. Xu. 2017. "Adaptable Blockchain-Based Systems: A Case Study for Product Traceability." *IEEE Software* 34 (6): 21–27. doi:10.1109/MS.2017.4121227.

Meijer, Albert, and Manuel Pedro Rodríguez Bolívar. 2016. "Governing the Smart City: A Review of the Literature on Smart Urban Governance." *International Review of Administrative Sciences* 82 (2): 392–408. doi:10.1177/0020852314564308.

Nakamoto, Satoshi. n.d. "Bitcoin: A Peer-to-Peer Electronic Cash System," 9.

Porru, Simone, Andrea Pinna, Michele Marchesi, and Roberto Tonelli. 2017. "Blockchain-Oriented Software Engineering: Challenges and New Directions." In *Proceedings of the 39th International Conference on Software Engineering Companion*, 169–171. ICSE-C '17. Piscataway, NJ, USA: IEEE Press. doi:10.1109/ICSE-C.2017.142.

Shen, C., and F. Pena-Mora. 2018. "Blockchain for Cities—A Systematic Literature Review." *IEEE Access* 6: 76787–76819. doi:10.1109/ACCESS.2018.2880744.

Uddin, M. A., A. Stranieri, I. Gondal, and V. Balasubramanian. 2018. "Continuous Patient Monitoring With a Patient Centric Agent: A Block Architecture." *IEEE Access* 6: 32700–32726. doi:10.1109/ACCESS.2018.2846779.

Wang, S., J. Wang, X. Wang, T. Qiu, Y. Yuan, L. Ouyang, Y. Guo, and F. Wang. 2018. "Blockchain-Powered Parallel Healthcare Systems Based on the ACP Approach." *IEEE Transactions on Computational Social Systems* 5 (4): 942–950. doi:10.1109/TCSS.2018.2865526.

Wessling, F., and V. Gruhn. 2018. "Engineering Software Architectures of Blockchain-Oriented Applications." In *2018 IEEE International Conference on Software Architecture Companion (ICSA-C)*, 45–46. doi:10.1109/ICSA-C.2018.00019.

Yuan, Y., and F. Wang. 2018. "Blockchain and Cryptocurrencies: Model, Techniques, and Applications." *IEEE Transactions on Systems, Man, and Cybernetics: Systems* 48 (9): 1421–1428. doi:10.1109/TSMC.2018.2854904.

Zhao, H., P. Bai, Y. Peng, and R. Xu. 2018. "Efficient Key Management Scheme for Health Blockchain." *CAAI Transactions on Intelligence Technology* 3 (2): 114–118. doi:10.1049/trit.2018.0014.

Chapter 10

Enabling Digital Twin through Blockchain: A Strategic Perspective

Ritika Wason, Broto Rauth Bhardwaj, and Vishal Jain

Contents

10.1 Introduction

In the 1990s the Internet paved the way for the digitization and global accessibility of information. However, this transfer of confidential data involved third parties ensuring secure transfer. As a result with passing time and increasing volumes of data, the secure transfer of assets was liquidated and somehow all personal data could be retrieved from the Internet either through ethical or unethical means [1]. Recent times have noted a rapid change in the world in terms of the rising digital transformation in almost all walks of life [2]. Industries have also noted a significant movement towards digital initiatives like the digital twin [3–6] to provide timely and quality services to the customer. This altered business model is actually an attempt to optimize operations to improve the overall customer experience. The new generation popularly known as the millennials are catalyzing this new trend. They prefer to use assets, products and processes as services rather than owning and maintaining them. It is estimated that by the turn of this decade the number of Internet of Things (IoT)-enabled devices will be over 20 billion across the globe [3]. These networked devices shall sustain millions of digital twins with important data. The digital twin technology amalgamates physical units with their virtually premeditated twins. It is actually mapping a physical entity with its digital image to enhance the performance and behavior of installed physical machines. It helps reduce downtime and increase performance. Digital twin technology is actually enabling the digital transformation of the entire society to facilitate competence and transparency in almost all endeavors of human life [2]. The digital twin is not a recent model, it has existed for several years and more lately has witnessed greater implementation from the perspective of machines where the digital copy of the machine (combining CAD/CAE technology, IoT and analytics) "lives" in conjunction with the physical machine and helps proactively foresee the need for repairs, potential improvement for enhanced performance, augmentation of the product line, etc. [7].

The IoT ecosystem demands IT components and connectivity. A digital twin can mirror a physical entity and facilitate the requisite digital supervision and maneuvering competencies of their physical counterparts in the real world. Furthermore they also allow efficient storage and transmission of data along with added services like analytics solutions which are key to domains like predictive modeling. Thus a digital twin can also prove to be an important component of realizing a smart city [7].

In recent years the blockchain technology has materialized into a novel organizing paradigm for the detection, evaluation and transmission of distinct units of all human activity at a greater magnitude than what has been achievable before. It is predicted that blockchains could significantly facilitate the application of digital twins in IoT [7]. The application of blockchain to digital twin is being termed as the rebirth of digital twin [7]. To truly leverage the benefits of digital twins in an IoT-enabled ecosystem a technology to transparently handle data-based transactions in a secure manner is desirable. Blockchain enables data exchange among multiple parties within a network, applying encrypted identities through peer-to-peer communication. The technology is expected to drive radical changes in varied sectors like enterprises, services and industries. It enables security through its built-in encryption mechanism while providing decentralized, transparent data access through immutable blockchain blocks. Hence the participating entities can trace past transactions any time through a secure and failure-free mechanism.

Though the digital twins have already been identified as a potential technology, as of now there is no universal, standardized mechanism implementing a digital twin. Currently, most information generated within the IoT ecosystem is stored in fragmented data silos. This fragmentation needs to be abolished and replaced by an integrated ledger technology, like the blockchain, in order to truly leverage the benefits provided by digital twins.

This chapter attempts to explore and analyze the integration of blockchains in enabling digital twins for realizing a smart IoT-enabled ecosystem. We aim to identify the main drivers and barriers of digital twins while exploring the possibilities of blockchain applications to digital twins. The next section elaborates upon how the blockchain technology can help realize digital twins. Section 10.3 evaluates certain real-time blockchain-enabled digital twins to assess the current situation. Section 10.4 delves into certain specific examples to completely understand and appreciate blockchain-enabled digital twins. Section 10.5 describes the challenges involved in unearthing the complete potential of a blockchain-enabled digital twin. Section 10.6 discusses how digital twins can help improve human life. Section 10.7 highlights some notable digital twin-enabled business trends. Section 10.8 concludes the chapter.

10.2 The Blockchain Perspective of Digital Twin

The blockchain technology is currently garnering lot of market expectations as well as interest due to its inherent characteristics like immutability, decentralization and time-stamped record storage [8]. Simply defined a blockchain is "a pooled, distributed, immutable ledger that assists the practice of recording transactions and trailing assets in a net of business connections" [9–11]. Initially developed by Nakamoto (2008) as a technology to support the Bitcoin cryptocurrency the technology has already demonstrated its potential for widespread adoption in varied industries [9, 12]. The technology is not just useful in moving capital; it can help in the transaction of any asset in a transparent, trustworthy way [13]. It creates an everlasting and transparent record of transactions while preventing intractable inhibitors across businesses [14]. Based on a peer-to-peer transmission operation network, transparent and cryptographically protected information facilitates smart contracts [15]. Rightly termed as the new paradigm for this decade, it holds the potential of realizing the connected world of computing through blockchain cryptography [16].

As per Deloitte and Austrian blockchain solution provider RIDDLE&CODE, "Blockchain technology is the most appropriate and competent means to create, examine and trade digital twins" [17]. In itself blockchain is an innovation that promises to be as disturbing as the Internet. The innovation enables decentralized, tamper-proof, community ownership of transaction records [18].

The connected world and industrial revolution 4.0 are also essential players along with the blockchain [19]. They actually imply multidevice computing that comprises wearable computing, laptops, smartphones, Internet of Things (IoT) sensors, smart homes, self-tracking devices (e.g. Fitbit) and smart cities. Blockchains can aid effective transfer of important information as well as efficient resource allocation in this connected world through smart contracts. These automatically accomplished contracts can perk up multiple operations in the public as well as private sector while eradicating the middleman [15]. Blockchains constitute a multifaceted platform capable of catering to diverse parties in a transparent and secure data transaction [18, 20]. It enables a trusted network between varied participating entities that share information to enable a transaction [21]. Through its indigenous features like a common, compatible protocol and data sharing through a common distributed ledger platform blockchain also enables interaction between all connected applications in different IoT ecosystems [11, 14, 15, 18, 22, 23]. To enable multistakeholder interactions among uniquely tokenized digital twins demands a comprehensible, competent platform with ample infrastructure facilities. Figure 10.1 depicts the main components of a blockchain.

Figure 10.1 depicts the primary characteristics of the blockchain technology. These characteristics help realize an immutable, irreversible transaction recorded

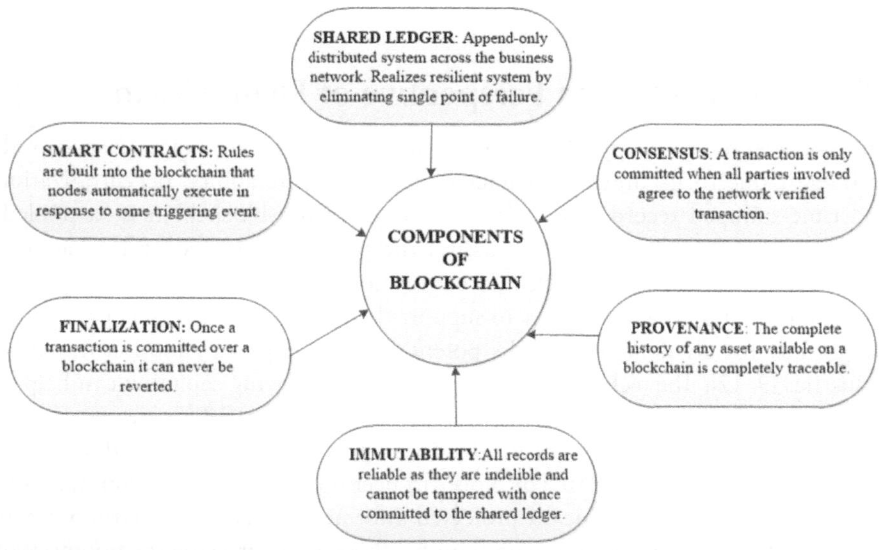

Figure 10.1 Prime characteristics of blockchain technology.

on a shared ledger permanently and simultaneously on each node of the concerned blockchain. This technology holds the potential to transform global economies by helping create and control digital replicas of all physical as well as logical assets, processes and frameworks as well as humans. A blockchain can serve as the backbone for such a system as it is an independent system without intermediaries. It is implemented through smart contracts in a peer-to-peer mechanism with built-in encryption as well as complete traceability of each data block. The technology has several inherent benefits like transaction transparency, fraud and manipulation control, corruption elimination, reduced costs, higher trust, lesser human errors, privacy, reliability and security among others [15, 24].

Though termed as the biggest invention since the Internet, complete harnessing of this technology is still hampered by a number of critical factors like lack of knowledge, empirical research, skilled developers and dilemmas due to absence of clear guidelines during implementation of smart contracts [15]. Many industries have already adopted the blockchain technology as a service to eliminate the barriers related to security as well as visibility in the transactional universe. Industries like real estate, insurance and finance, healthcare, retail, defense, transport, agriculture, etc., have already adopted the blockchain to instill trust as well as security in their transactions. We elaborate further on some considerable blockchain use cases in the following sections.

10.3 Real-Time Blockchain-Enabled Digital Twins

Since 2017 many industries have shown interest in developing blockchain-enabled digital twins. We introduce a few projects in Table 10.1 to illustrate the opportunities and challenges involved.

Table 10.1 proves beyond doubt that digital twins are the main constituents in the present digital economy that are helping realize a digital society. However, it is important to understand that to create a digital data imprint of any social, economic or human asset, entity or process it is important to have an automated, distributed, failure-free, smart, trustable, irreversible transaction as well as data-lodging mechanism. This requirement is fulfilled by blockchain technology. Hence, blockchain-enabled digital twins are the future digital twins or digital replicas for anything and everything. Instead, it is this technology that has enabled the conjoining of existing digital, physical and social spheres [46].

10.4 Some Notable Cases

Digital twins are gaining momentum as they offer real-time transparency [47]. Applied in many domains like aviation, manufacturing, automobile, etc., it is appropriately referred to as game-changing technology. In this section we attempt

Table 10.1 Notable Digital Twin Initiatives

S. No	*Blockchain-Enabled Digital Twin Project*	*Company*	*Description*	*Challenges Ahead*
1.	**S/4HANA Cloud for Intelligent Product Design** [4, 25]	SAP	Cloud solution on cloud SAP platform. It aids in managing and digitalizing your product research and development. Storage and sharing as well as review of engineering documents are fostered through a process-driven approach.	To maintain compliance with data protection regulations, industry-specific legislations in different countries especially in personal data deletion scenario.
2.	**Energy metering** [26]	Siemens, Emerson	Siemens Electrical Digital Twin initiative closely aligns real and virtual worlds by providing utilities with a single synchronization point to model data across their IT landscape. This digital twin replica of the live plant runs in parallel to the real control system enabling advanced testing and ensuring unaltered electric supply.	Digital twinning can be applied to every energy generation or distribution site; however uninterrupted data access remains the primary issue to ensure successful twin operation.

(*Continued*)

Table 10.1 (Continued) Notable Digital Twin Initiatives

S. No	*Blockchain-Enabled Digital Twin Project*	*Company*	*Description*	*Challenges Ahead*
3.	**GE Digital** [27] **Predix Platform**	General Electric	GE Digital is a subordinate of the American multinational conglomerate corporation General Electric. GE Digital functions across a number of industries, including aviation, manufacturing, mining, oil and gas, power generation and distribution, and transportation among others. GE Digital defines a hierarchy of digital twins like component twin, asset twin, system twin and process twin. These varied twins effectively help to monitor, simulate and control online or offline assets or processes.	Globalization, novel manufacturing techniques and liberalization policies are the potential challenges. Further managing all the design data for digital twin amongst collaborators and suppliers while the physical products change shall be a test.

(Continued)

Table 10.1 (Continued) Notable Digital Twin Initiatives

S. No	*Blockchain-Enabled Digital Twin Project*	*Company*	*Description*	*Challenges Ahead*
4.	**Aviation** [2, 28, 29]	Boeing	They are creating a digital twin of each plane they are manufacturing in order to keep it up to date in real time. This helps airlines optimize their data efficiency. Helps in enhancing customer value through services like immediate remote assistance through a HoloLens and Skype. Further it is allowing the aviation industry save space through prescriptive and predictive analytics.	Availability of suitably complete and accessible digital data for all equipment is the main challenge for the aviation digitization as much of the equipment in use in the aviation sector was put into use long ago using older technologies, hence information relating to such components may be missing, incomplete or incorrect. This combined with challenges related to data security, ownership, volume and integrity need to be efficiently catered.
5.	**Connected cars** [2]	Ford	Digital twins are being successfully applied in the automobile sector to generate the virtual model of a connected vehicle. This twin helps continuously analyze vehicle performance as well as those of the connected components. This helps in enabling a customized service for each customer.	The right kind of probes so that behavioral data is captured and continuously updates the virtual vehicle. The instrumentation is tricky since it has been performed without compromising the system performance.

(*Continued*)

Table 10.1 (Continued) Notable Digital Twin Initiatives

S. No	*Blockchain-Enabled Digital Twin Project*	*Company*	*Description*	*Challenges Ahead*
6.	**Steel manufacturers** [30]	POSCO	Are using digital twins to simulate their complex manufacturing operations, empowering them to predict issues far in advance even before processes and products reach the factory floor. They have achieved this digitization through Dassault Systèmes' 3DEXPERIENCE platform.	To unlock the actual value of the digital twin demands a holistic approach to accumulate, control and direct the digital data of the product. A robust engineering change management process is also desirable to guarantee that the digital twin precisely manages the virtual and physical configurations.
7	**Housing development** [31]	Global Housing Builders	Enables a transparent housing development marketplace with more prospects for the less recognized firms to participate.	Common, holistic policies and regulations across the global marketplace are crucial for the successful implementation of this trend.
8	**Maritime industry** [8, 32, 33]	Norway	Blockchain is being effectively applied to reduce the pollution generated by the maritime industry.	Establishing successful interoperability between smart space components is crucial for effective implementation.

(Continued)

Table 10.1 (Continued) Notable Digital Twin Initiatives

S. No	*Blockchain-Enabled Digital Twin Project*	*Company*	*Description*	*Challenges Ahead*
9	**Secure plug and produce** [34]	Asset Administration Shell (AAS)	The concept expects that as soon as a new module is connected to a system, the data configuration transfer begins. Combining AAS with blockchain ensures uniform, standardized, authentic transfer of configuration data.	All system components require successful behavior adaptation to imbibe self-x capabilities like self-optimization, auto configuration, etc.
10	**Edge marketplaces** [35]	Europe	These marketplaces enable support to multiple providers for offering services at the network edge.	Resilience, cost, and quality of service and experience will subsequently enable such a marketplace to adapt its services over time.
11	**Traffic congestion avoidance** [36]	Deloitte	Proposes real-time, flexible, precision-centric and predictive traffic monitoring, quantity, administration and enrichment resolutions towards sustainable smart cities.	Delivering real-time road status specific alerts shall require effective and digitally connected, smart mobile apps for efficient service delivery.

(*Continued*)

Table 10.1 (Continued) Notable Digital Twin Initiatives

S. No	*Blockchain-Enabled Digital Twin Project*	*Company*	*Description*	*Challenges Ahead*
12	**Individual competence portfolio authorization** [37, 38]	Russia	Masterchain platform developed as a distributed, reliable and flexible system of records of events like diploma issuance to harmonize enterprise expectations with individual interests.	Proper ontologies shall be required to ensure successful operation of the twin.
13	**Space-based digital twins** [39, 40]		Recent lighter and cheaper nanosats are being deployed to generate new services as well as supply chains in space. This can enable creation of an immutable, trusted, digital twin of the earth. This replica could be utilized for almost anything and everything.	Regulating this infrastructure to prevent its misuse shall be of utmost importance in its actual realization.
14	**5G network slice broker** [41]	5th Generation Mobile Networks	The notion aims to enable mobile virtual network operators, over the-top providers as well as industry players to request and lease resources as and when required. It involves use of the Blockchain Slice Leasing Ledger Concept to realize Factory of the Future.	Next-generation mobile broadband requires wireless performance, privacy and security lacking in existing technologies.

(*Continued*)

Table 10.1 (Continued) Notable Digital Twin Initiatives

S. No	*Blockchain-Enabled Digital Twin Project*	*Company*	*Description*	*Challenges Ahead*
15	**e-Village**[42]	India	Government facilities and services require citizen essential information. A government trusted information repository can be created by digitizing the parivar register through blockchain. This shall enable quicker service delivery to one and all.	All geographically distinct villages need to be connected through ICT to realize this model.
16	**Global FinTech revolution** [43]	China, India	Digitized financial technology services linked to mobiles have generated the financial revolution capturing the normal as well as the financially underserved, i.e. the banked and unbanked consumer alike. China has emerged as the leader in this FinTech revolution while India has materialized as a massive testing ground for financial inclusion and innovation.	In a demographically diverse country like India three key enablers often referred to as the JAM trinity including Jan Dhan Yojana, Aadhar and Mobile phones have paved the way for multiple novel technologies and services.
17	**3D printing** [21]	Moog Aircraft Group	Applied to ensure secure 3D printing of aircraft components via blockchain.	Automation of creation and testing as well as servicing of all such aircraft components shall need effort.

(Continued)

Table 10.1 (Continued) Notable Digital Twin Initiatives

S. No	*Blockchain-Enabled Digital Twin Project*	*Company*	*Description*	*Challenges Ahead*
18	**Supply chain management** [21, 44]	IBM	Blockchain is successfully applied for tracking of containers during shipping, register certifications and crucial product information through the entire chain.	Digitization of the complete shipping supply chain involving many conventional components is a challenge.
19	**Education** [21]	Holburton School, San Francisco	This school is applying blockchain to store and deliver its certificates and curb fake certificate issue.	Globally empowering all universities and schools to adopt this technology though challenging shall help eliminate many scams in education.
20	**Cyber-physical systems (CPSs)** [45]	DeCyMo	Enable smart monitoring and control in industrial as well as home scenarios. May help in monitoring energy consumption, asset management, etc.	Large-scale deployments of the solution require testing to test scalability as well as resilience.

to understand and appreciate the knowledge, assets, data and intelligence that go into realizing a successful digital twin. We discuss the case of a smart factory, the digital version of the crucial supply chain manufacturing process [7, 48–51]. The progression of creating a digital twin usually follows this path:

i. Understanding the operation and prediction at the asset level and leverage to optimize individual performance.
ii. Optimize maintenance at individual level.
iii. Aggregate for multiple assets and optimize them at the operations level.
iv. Rethink business models and enable new values and services.

Connectivity is crucial to the manufacturing process. With the rise of industry 4.0 and the confluence of digital and physical worlds, supply chain dynamics has undergone complete transformation. The new open chain mechanism of supply chain operations better known as the digital supply network is actually the basis for future competition in the manufacturing industry. However, to completely comprehend these, digital supply network firms need to provide parallel integration through the innumerable operational systems that influence the business; vertical integration through the associated manufacturing systems; and end-to-end, holistic amalgamation through the entire value chain. Together this integration is known as the smart factory that enables greater value both within the factory and across the supply network. The smart factory enables a fully connected, flexible system that utilizes a constant stream of data from connected operations and production systems to adapt to new demands through the digital twin. The resultant is a flexible, agile system with higher efficiency and reduced downtime. Such a system can self-optimize performance across a broader network, self-adapt to and learn from dynamic conditions in real-time and autonomously run the entire production process. Hence, such a system can evolve and grow with the changing needs of the organization. Figure 10.2 displays the features of a smart factory that are critical to its success.

As displayed in Figure 10.2 connectivity, optimization, transparency, proactivity and agility are the major attributes of a smart factory. Together they ensure better decision making, helping organizations improve their production process.

In the Indian perspective, Hero Moto Corp, the world's largest two-wheeler manufacturer, is the first automobile company in India to have adopted the digital twin approach in April, 2016 [52]. This demanded collaboration across varied departments, to adopt a digitally enabled manufacturing facility to achieve reduced costs and better quality. They have realized a digital replica for their Vadodara (Halol) manufacturing facility to digitally visualize the facility continuously to enable timely changes and enhancements.

The company aims to visualize its product and process as well as resources in a virtual context to enhance productivity, reduce costs and eliminate disruptions for each of its manufacturing facilities. The company also achieved pre-validation of

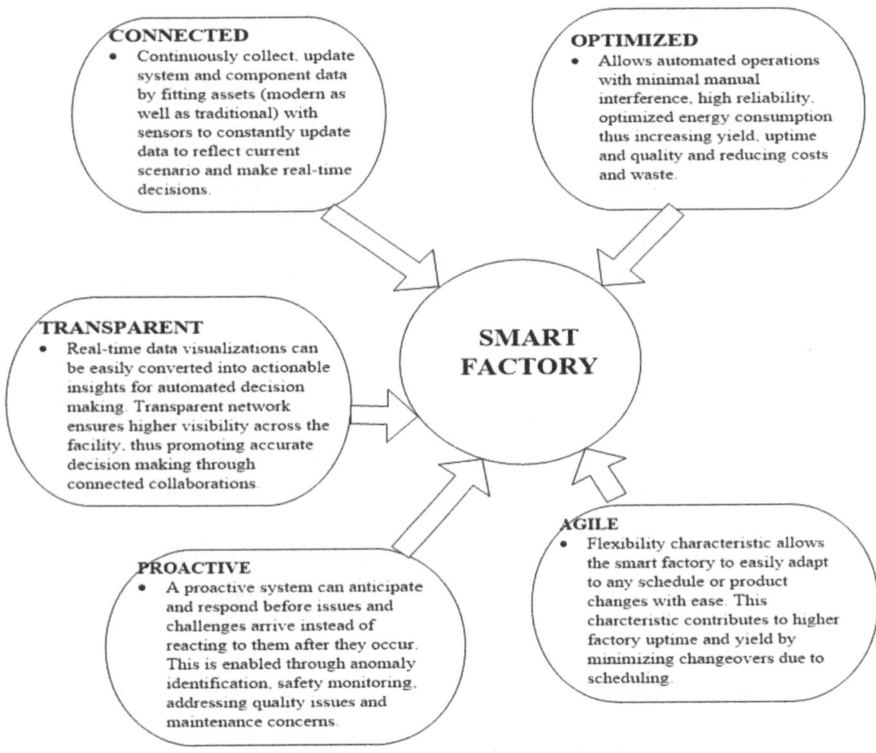

Figure 10.2 Key attributes of a smart factory.

each of its newly commissioned manufacturing facilities, prior to actual physical commissioning to eliminate associated costly reworks. The process planning for existing and new models on the assembly line was also validated to achieve process standardization and a single, common repository for all data. Further the company was successful in achieving cost and time reduction at varied stages of development and manufacturing along with the availability of 3D training material to achieve paperless shops.

10.5 Challenges to Realizing Blockchain-Enabled Digital Twins

Swift distribution of digital replicas faces snags due to the lack of semantic interoperability between architectures, values and ontologies. The expertise essential for automatic detection is also in short supply [47]. Though blockchains are being projected as the next big thing facilitating the evolution of digital twin and IoT, there are still some challenges to be answered, like [53]:

i. How shall a blockchain provide the "**right to be forgotten**" [54]?: An inherent characteristic of the blockchain is that it is practically impossible for the data to be deleted. In such a case how shall a blockchain enable the right to forget data in a digital twin? A proposed solution is through usage of a unique root key for users. This root key could be applied to generate a new key for each individual transaction. As a result each transaction could have different hashes and no longer be associated to each other.

ii. **Informed and clear-cut policies** [55]: Adopting a blockchain-enabled digital twin would imply a complete paradigm shift for the participant organization. For this the organization requires proper awareness regarding its role in the IoT-enabled ecosystem, what role would they like to play (data provider or data acquirer) and appropriate details of the specific blockchain infrastructure required to achieve desired level of cross-industry efficiency. Appropriate technical as well as hardware and software skills are required to embrace this change. Industries, professionals and businesses are yet to fully adapt themselves to these changing dynamics.

iii. There can **be no generic twin** [55, 56]. This simply implies that one twin shall not fit all. Thus one needs a digital twin for each product that one manufactures. This is especially true for all products involving human interaction.

iv. **Digitizing everything** [57]: To realize the true potential of digital twins and make them as popular and omnipresent as the Internet it is required to digitize all data. This implies that even small and medium-scale manufacturers or service providers need to initiate the process towards digitization.

v. **Efficient product development platforms and processes** [55, 56]: To create efficient digital replicas requires appropriate product development platforms as well as processes. However, these technologies are still in their nascent stages. Thus, as per Marc Halpern, "digital twins are still more about visions and promises rather than finished solutions".

vi. **Securing digital twins** [57]: With passing time, digital twins shall themselves become intellectual capital. Ownership, security and other issues related to this intellectual capital require serious thought. One should also not forget that these IoT devices shall also be vulnerable to distributed-denial-of-service (DDoS) attacks, hacking and data theft as well as data hijacking issues in real life. How to overcome these issues is another significant challenge.

vii. **Proper regulations** [58]: To match the technological developments one requires proper regulations related to account of ownership and transactions.

viii. **Reliability** [59]: Blockchain-enabled digital twins demand reliable, 24/7, ubiquitous connectivity under all environments to ensure optimal operation in all lifecycle stages. This shall be challenging where assets are deployed under harsh terrestrial domains.

ix. **Resilience**: Wireless networking resilience will need to be improved massively, enabling continuous uptime and fastest recovery, in turn creating an "always on" digital ecosystem.

x. **Interaction between smart components within and with other** [32]: This can be the primary challenge in successful application of blockchain-enabled digital twins in varied domains. Successful interaction between all components of a twin is necessary to ensure accurate replication of the physical asset be it human or physical entity. Hence, establishing healthy, continuous interaction between all smart components of a twin as well as with other twins as well as entities in the ecosystem is essential for the effective implementation of blockchain-enabled digital twins across the industry.

The above challenges are just the tip of the iceberg, there are many more challenges and issues that shall keep coming up as digital twins are applied across varied businesses and domains.

10.6 Digital Twin in the Human Context

Just like the digital twins are benefiting the effectiveness and efficiency of assets in the industry, they can also be beneficial for the human health and life sciences industry [60]. To realize this we first need to understand that as humans we all are creating footprints of ourselves on the Internet. All information about ourselves that we leave on the Internet in any form (attributes, actions, online presence) is actually creating a digital twin for us. This data footprint that we are creating on the Internet if connected with advances in IoT and analytics can help monitor and predict our future needs especially in healthcare and otherwise also [61]. Our LinkedIn, Facebook, Twitter, Snapchat and Instagram IDs are actually our digital replicas easily available on the Internet. To further appreciate this, we first evaluate the components that will help realize a human digital twin [60, 62, 63].

i. **Attributes**: They are the core data that constitute our identity. For example: Name, age, gender, address, education, nationality, etc. These data help map an individual and predict whether they are prone to certain health conditions due to origin or dwelling.

ii. **Interactions**: This footprint is generated on the basis of our interaction with the external world. For example: Impact of frequent air travel on one's health, shopping habits' relation to bank data. New technology like Fitbit watches, interaction data with doctor, phone usage statistics, etc., are all essential components of the "Digital Me" that if consolidated with attributes can aid in monitoring, diagnostics and prognostics of individual performance and well-being. These data may even help organizations in their decision making. However, these decisions involve a lot of issues we would not discuss here.

iii. **Online persona**: With the Internet, each one of us also has a digital identity through which we do all browsing. These data are already stored and shared by varied websites and can again provide interesting insights into many of our personal attributes like health, interests, etc.

The above three attributes clarify the possibility of the human digital twin. Just like a machine digital twin can aid in gauging the need for maintenance, growth, etc., human digital twins can help in varied life-savvy predictions.

Clearly, the above discussion clarifies the advantages of digital twins for humans. However, if such a digital replica is built on traditional technologies, it will hold all data in a central repository which would require guarding along with analytics to provide guarded predictions. To overcome the additional security costs blockchains provide a viable alternative. Blockchains enable the creation of identities on an immutable ledger with no centralized owner. Such self-sovereign identity is capable of deciding who gets access to their data and can also track all those who accessed it. For example: The Hyperledger Indy blockchain platform enables the creation of such identities. Such a blockchain network is generally constituted by the data owner (individual), the data attester (university that issued the qualification certificates) and the data requestor, say the employer. Governance of such a network can be democratic and each access to data can be authorized by the data owner. This ability to create digital replicas of citizens holds the potential to transform the global economy. Many governments across the globe are investing large amounts in blockchain-enabled initiatives like self-sovereign identities, ownership registration for movable assets, verification, etc. Blockchain-enabled human digital twins are going to be a vital component in realizing and implementing all such initiatives [64]. Table 10.2 lists some notable government initiatives realizing the same.

Table 10.2 depicts how different governments across the globe have realized the importance of digital twins and are spending huge capital to realize blockchain-enabled twins to digitize and improve citizen life in different scenarios [44, 74]. What is notable from Table 10.2 is that varied governments or private players in different countries have simply embraced the blockchain technology to provide safer, easier and better-connected citizen services. This simple adoption of blockchain in processes has eliminated the need for an intermediary third party, risk of data leakage, etc., while enabling services for one and all alike. Further these blockchain-powered processes are generating human digital replicas in the background while improving human quality of life.

10.7 Digital Twin-Enabled Business Trends

The Internet of Things (IoT) has enabled network connections to ease communication between devices and methods. Companies have been deploying smart devices in many new ways to enhance their businesses. Here are four trends we can look forward to [19, 75, 76]:

Table 10.2 Government Blockchain-Enabled Digital Twin Initiatives

S. No	*Government*	*Project Name*	*Description*
1	**India**	IndiaChain [65]	The Indian Government through its think tank NITI Aayog is currently working on an official blockchain solution to enable digital, tamper-proof degree certificates for 2019 students of IIT, Bombay and varied colleges of Delhi University to start with.
2	**State: Andhra Pradesh Country: India**	FinTech Valley Vizag [66]	In 2016, Andhra Pradesh became the first Indian state to implement blockchain for governance. It has piloted two main projects: Managing land records and streamlining vehicle registrations while building Vishakhapatnam into a world-class ecosystem while collaborating with government, academia, corporate, investors and entrepreneurs.
3	**Estonia**	E-Estonia [67]	The Estonian Government through digital twins has succeeded in building the world's most advanced digital society that started with the introduction of digital service tax as early as 2000 to the extent that 99% of its public services are available online.
4	**Canada**	Concierge service [68]	Creates a digital duplicate of a physical construction site that enables project teams to be more certain in their decision-making and deliver projects more efficiently, from design to construction.

(*Continued*)

Table 10.2 (Continued) Government Blockchain-Enabled Digital Twin Initiatives

S. No	*Government*	*Project Name*	*Description*
5	**Russia**	e-Government [69]	Enabling digital services in parallel with other channels. The progress includes implementation of the Multi-Function Centres and a Unified Portal; setting up infrastructure to link different government institutions; establishing national databases; and introducing common services such as identification, authentication and payments systems.
6	**European Union**	DigiTwins [47]	DigiTwins, a huge research scheme in Europe and beyond, plans at transforming healthcare and biomedical research for the assistance of citizens and society through the creation of digital twins. These twins are precise computer models of the key biological practices within each individual that keep us healthy or lead to disease. They can be used to recognize individually finest therapies as well as preventive and lifestyle measures, without exposing individuals to unnecessary risks or healthcare systems to unnecessary costs.
7	**Indonesia**	Smart City [70]	All metropolitan and big cities in Indonesia are planned to be made smart cities by 2045. To realize the above a properly designed master plan detailing all city requirements as well as technology requirements will serve as a pre-requisite. A resilient, properly planned IT framework is crucial for the successful realization of a smart city.

(*Continued*)

Table 10.2 (Continued) Government Blockchain-Enabled Digital Twin Initiatives

S. No	*Government*	*Project Name*	*Description*
8	**Netherlands**	Smart Industry Field Labs [71]	Dutch smart industry field labs are being used to accelerate digitization of the industry. They are basically public private partnerships that work to develop, test and implement smart industry solutions. Over 72 million euros have been invested in such labs since 2015 through public as well as private financing.
9	**China**	Alibaba's Rural Taobao Strategy [72]	China's rapid growth in the last four decades has also noted its rapid transition from a rural to an urban society. However, China has also ensured timely e-commerce and digital financial services access to its existing rural population as early as 2014 through Alibaba Group, rural Taobao Strategy.
10	**Kenya**	M-Pesa FinTech Revolution [43]	In March, 2007 the M-Pesa electronic money transfer project was launched in Kenya by Vodafone and Safaricom enabling users to store money in their mobile phone SIM cards as electronic currency and utilize the same for varied transactions and services.
11	**United Kingdom (UK)**	Digital Construction [73]	The UK government is trying to foster digital construction by launching degree apprenticeships, a novel model of undergraduate education that could in turn leverage digital capability in the construction industry.

10.7.1 Constant Consumer Commitment

Smart products enable users to communicate with and control appliances as well as other household products through IoT. These communication capabilities can also be applied for monitoring products and enabling proactive support.

For example, if some car part requires servicing or otherwise fails, a message can be transmitted to the customer as well as the manufacturer's service system.

10.7.2 Business Process Supervision

Everything quantifiable can be improved. IoT shall further enable detailed monitoring in other domains of business, like office work and field processes.

Wearable devices are sensors that can be applied to individuals to scrutinize activities and record information. These sensors and what they communicate can convey more data to offer deep analytics which can augment productivity and help lower costs.

10.7.3 Automated Services

Varied industries including automobile can now deploy IoT sensors on vehicles and packages to maximize the visibility of supply chains and improve operations of valuable assets. These sensors can identify variations in temperature, light and other attributes, and can be used to alleviate risks of delays, interruption, robbery and more.

Real-time alerts can enable services to be at the right place at the right scheduled delivery times, and with the apparatus required to amplify efficiencies.

10.7.4 Expanded Big Data

Sensors deployed as part of IoT devices fabricate data that can be stored and analyzed as part of existing analytics – generating completely new classes of findings.

10.7.5 Embedded IT

The transition from accumulation of IoT "smarts" to current devices and generating devices with IoT components built-in will be transformative. However this big industrial transformation rightly referred to as Industrial Revolution 4.0 shall actually transform the industrial landscape and its mechanisms.

10.8 Conclusions

The Internet of Things and digital twins as well as blockchains have been successfully implemented in varied industrial sectors and domains. Together they are instrumental in realizing the new industry 4.0 revolution. However their implementation has always faced technical as well as financial limitations. As the number and volume of IoT devices increase, the technical and financial challenges diminish. However, for the relatively novel digital twin platform we need to realize that they are just not an academic modeling exercise. Digital twins are created for specific outcomes or what may be termed as key performance indicators (KPI). These KPIs are integral to deciding issues like maintaining a specific quality of service, predicting component lives to reduce downtime, etc. However, an integrated, stable underlying platform was missing till now. The underlying distributed ledger technology of a blockchain offers immutability and transparency as well security benefits for its participating entities. Hence, blockchain has emerged as a promising underlying platform to realize an effective and efficient digital twin. Instead it can be appropriately termed that digital twins are the technology of the future which hold the potential of transforming corporate as well as human life.

10.9 Future Scope

The Digital Revolution is surely sweeping across the globe, transforming mankind and industries as well as economies alike [43]. Everything around us, be it assets, solutions, processes or even technologies themselves, is transforming into services. Blockchain-enabled digital twins are surely an innovative technology that holds the capability of revolutionizing the industrial landscape as well as transforming human life. However, there is still a long way to go before realizing the complete potential of this technology as depicted by the research and academia. Many future potential areas of research and effort can be understood and outlined from this study. However, they are not complete and may keep expanding with novel applications of this technology to varied domains:

1. **Smart contract ontology development** [31]: Ontologies can be developed for the description of smart contracts to provide interoperability for the interaction between smart components as well as smart assets. This may help in finding solutions to the problems arising in blockchains.
2. **IoT platform shortcomings** [32]: A detailed analysis of existing IoT platforms available for the establishment of blockchain-enabled digital twins and smart assets suggests that they still suffer from a number of shortcomings, like lack of authorship verification mechanisms, information durability issues, effective control over resource exchange in production, etc. These issues are still open to effective, persistent solutions.
3. **Block immutability shortcomings** [59]: This feature of the blockchain is a boon as well as a bane. It should be noted that the constantly expanding chains of blocks shall consistently demand potential chunks of memory which may not be available with some seemingly simple devices operating

in the IoT ecosystem. To solve these issues again effective ontologies may be used to delegate some information of the weak devices over to some strong devices in the chain.

4. **Bigger and bigger big data** [77]: This new wave of digital transformation includes digitalization as well as digitization. Digitization involves creating digital replicas of physical assets, entities, etc., while digitalization involves the use of appropriate technologies to effectively use digitized entities in societal processes. This complete process generates the creation and usage of bigger and bigger amounts of big data that require effective mechanisms of storage, analysis and sharing.

References

1. M. Miscevic, Gea Tijan, Edvard Zgaljic, Drazen Jardas, "Emerging trends in e-logistics," *Mipro*, pp. 1353–1358, 2018.
2. S. Nadella, J. Euchner, "Navigating digital transformation," *Res. Manag.*, vol. 61, no. 4, pp. 11–15, 2018.
3. Deloitte and Riddle& Code, "IoT powered by Blockchain How Blockchains facilitate the application of digital twins in IoT," p. 20, 2018.
4. J. Kobielus, "Networked digital twins are coming to industrial blockchains - SiliconANGLE," 2018. [Online]. Available: https://siliconangle.com/2018/04/24/networked-digital-twins-coming-industrial-blockchains/. [Accessed: 07-Sep-2018].
5. T. Rueckert, "Digital twin + blockchain - SAP news center," 2017. [Online]. Available: https://news.sap.com/2017/05/sapphire-now-digital-twin-blockchain/. [Accessed: 07-Sep-2018].
6. Gary Schwartz, "Our digital twin & the blockchain – Gary Schwartz," 2018. [Online]. Available: https://www.ifthingscouldspeak.com/2018/05/14/our-digital-tw in-the-blockchain/. [Accessed: 07-Sep-2018].
7. S. Wang, J. Wan, D. Li, C. Zhang, "Implementing smart factory of industrie 4.0: An outlook," *Int. J. Distrib. Sens. Networks*, vol. 2016, 2016.
8. A. H. Gausdal, K. V. Czachorowski, M. Z. Solesvik, "Applying blockchain technology: Evidence from norwegian companies," *Sustain*, vol. 10, no. 6, pp. 1–16, 2018.
9. S. Nakamoto, "Bitcoin: A peer-to-peer electronic cash system," *Www.Bitcoin.Org*, p. 9, 2008.
10. J. Yli-Huumo, D. Ko, S. Choi, S. Park, K. Smolander, "Where is current research on Blockchain technology? - A systematic review," *PLoS One*, vol. 11, no. 10, pp. 1–27, 2016.
11. P. Deepak, M. Nisha, S. P. Mohanty, "Everything you wanted to know about the blockchain," *IEEE Consum. Electron. Mag.*, vol. 7, no. 4, pp. 6–14, 2018.
12. D. Folkinshteyn, "A tale of twin tech: Bitcoin and the www," *J. Strateg. Int. Stud.*, vol. X, no. 2, pp. 82–90, 2015.
13. M. Weeks, "The evolution and design of digital economies," 2018.
14. B. Pellot, "Fast forward," *Ibm*, vol. 42, no. 3, pp. 46–49, 2013.
15. P. A. Corten, "Blockchain technology for governmental services: Dilemmas in the application of design principles," pp. 1–14, 2017.
16. M. Swan, *Blockchain: Blueprint for a New Economy*, 2015, O'Reilly Media: Newton, MA.

17. D. Schahinian, "IoT forecast: Digital twins to be combined with blockchain - digital twin - HANNOVER MESSE," 2018. [Online]. Available: https://www.hannovermesse.de/en/news/iot-forecast-digital-twins-to-be-combined-with-blockchain-88960.xhtml. [Accessed: 29-Nov-2018].
18. Z. Zheng, S. Xie, H. Dai, X. Chen, H. Wang, "An overview of blockchain technology: Architecture, consensus, and future trends," In *Proc. - 2017 IEEE 6th Int. Congr. Big Data, BigData Congr. 2017*, pp. 557–564, 2017.
19. B. Cearley Walker, "Top 10 strategic technology trends for 2017," no. October 2017, 2016.
20. Z. Zheng, S. Xie, H.-N. Dai, X. Chen, H. Wang, "Blockchain challenges and opportunities: A survey Shaoan Xie Hong-Ning Dai Huaimin Wang," *Int. J. Web Grid Serv.*, vol., 14, no. 4, pp. 1–24, 2016.
21. Virbahu Nandishwar Jain, Devesh Mishra, "Blockchain for supply chain and manufacturing industries and future it holds!" *Int. J. Eng. Res.*, vol. 7, no. 9, pp. 32–40, 2018.
22. G. Chen, B. Xu, M. Lu, N.-S. Chen, "Exploring blockchain technology and its potential applications for education," *Smart Learn. Environ.*, vol. 5, no. 1, p. 1, 2018.
23. L. Lee, "New kids on the blockchain: How bitcoin's technology could reinvent the stock market," *Hast. Bus. Law J.*, vol., 12, no. 2, pp. 81–132, 2015.
24. A. Bahga, V. K. Madisetti, "Blockchain platform for industrial internet of things," *J. Softw. Eng. Appl.*, vol. 9, no. 10, pp. 533–546, 2016.
25. "Introduction - SAP help portal." [Online]. Available: https://help.sap.com/viewer/d3a4810ff9dd41c59c50e1d1a6d4d7ae/1811/en-US. [Accessed: 19-Nov-2018].
26. "Prepare for a sustainable digital future Enable interoperable data exchange and synchronization."
27. "Digital twin | GE digital." [Online]. Available: https://www.ge.com/digital/applications/digital-twin. [Accessed: 19-Nov-2018].
28. S. Datta, "Cybersecurity-an agents based approach?" 2017.
29. "Digital twins and threads in aviation, aerospace and defense," 2017. [Online]. Available: https://www.capgemini.com/us-en/2017/12/establishing-a-fully-functional-digital-twin-or-digital-thread-in-aviation-aerospace-and-defense/. [Accessed: 19-Nov-2018].
30. S. Dudley, "Dassault systèmes helps POSCO digitise its manufacturing operations," 2015. [Online]. Available: http://www.technologyrecord.com/Article/dassault-syst232mes-helps-posco-digitise-its-manufacturing-operations-49171. [Accessed: 30-Nov-2018].
31. I. Savu, G. Carutasu, C. L. Popa, C. E. Cotet, "Quality assurance framework for new property development: A decentralized blockchain solution for the smart cities of the future," *Res. Sci. Today*, vol. 13, 2017.
32. N. Teslya, I. Ryabchikov, "Blockchain-based platform architecture for industrial IoT," In *Conf. Open Innov. Assoc. Fruct*, pp. 321–329, 2018.
33. K. Czachorowski, M. Solesvik, Y. Kondratenko, *The Application of Blockchain Technology in the Maritime Industry*, vol. 171. Springer International Publishing, 2019.
34. D. Lang, M. Friesen, M. Ehrlich, L. Wisniewski, J. Jasperneite, "Pursuing the vision of industrie 4.0: Secure plug-and-produce by means of the asset administration shell and blockchain technology," In *2018 IEEE 16th International Conference on Industrial Informatics (INDIN)*, pp. 1092–1097, 2018.
35. B. Varghese, M. Villari, O. Rana, P. James, T. Shah, M. Fazio, R. Ranjan, "Realizing edge marketplaces: Challenges and opportunities," *IEEE Cloud Comput.*, vol. 5, no. 6, pp. 9–20, 2018.

36. S. A. P. Kumar, R. Madhumathi, P. R. Chelliah, L. Tao, S. Wang, "A novel digital twin-centric approach for driver intention prediction and traffic congestion avoidance," *J. Reliab. Intell. Environ.*, vol. 4, no. 4, pp. 199–209, 2018.
37. F. Wei, N. N. Pokrovskaia, "Digitizing of regulative mechanisms on the master-chain platform for the individualized competence portfolio," In *2017 IEEE VI Forum Strategic Partnership of Universities and Enterprises of Hi-Tech Branches (Science. Education. Innovations) (SPUE)*, pp. 73–76, 2017.
38. G. A. Kostin, N. N. Pokrovskaia, M. U. Ababkova, "Master-chain as an intellectual governing system for producing and transfer of knowledge," In *2017 IEEE II International Conference on Control in Technical Systems (CTS)*, pp. 71–74, 2017.
39. F. Longo, "Advanced data management on Distributed Ledgers: Design and implementation of a Telegram BOT as a front end for a IOTA cryptocurrency wallet," July 2018.
40. "Blockchain and space-based 'Digital Twin' of earth | Bitcoin Magazine," 2017. [Online]. Available: https://bitcoinmagazine.com/articles/blockchain-and-space-based-digital-twin-earth-offer-insights-and-web-connectivity/. [Accessed: 06-Dec-2018].
41. J. Backman, S. Yrjola, K. Valtanen, O. Mammela, "Blockchain network slice broker in 5G: Slice leasing in factory of the future use case," In *Jt. 13th CTTE 10th C. Conf. Internet Things - Bus. Model. Users, Networks*, vol. 2018–January, pp. 1–8, 2018.
42. D. Faizan, S. Ishrat, "Impeccable renaissance approach: An e-village initiative," In *ICACCT 2018*, vol. 899, pp. 335–346, 2018.
43. I. Limited, "Services in the time of being digital," *Infosys Insight*, vol. 4, no. 1–104, 2016.
44. P. Jiang, "Social manufacturing paradigm: Concepts, architecture and key enabled technologies," *Adv. Manuf.*, pp. 13–50, 2019.
45. P. Gallo, U. Q. Nguyen, G. Barone, P. van Hien, "DeCyMo: Decentralized cyber-physical system for monitoring and controlling industries and homes," In *2018 IEEE 4th Int. Forum Res. Technol. Soc. Ind.*, pp. 1–4, 2018.
46. R. N. Bolton, J. R. McColl-Kennedy, L. Cheung, A. Gallan, C. Orsingher, L. Witell, M. Zaki, "Customer experience challenges: Bringing together digital, physical and social realms," *J. Serv. Manag.*, vol. 29, no. 5, pp. 776–808, 2018.
47. S. Datta, "Emergence of digital twins is this the march of reason?" *J. Innov. Manag.*, vol. 5, no. 3, pp. 14–33, 2017.
48. D. Communications, "The smart factory," *Complete Networked Value Chain.*
49. B. Sniderman, M. Monika, M. J. Cotteleer, "Industry 4.0 and manufacturing ecosystems: Exploring the world of connected enterprises," Deloitte University Press, pp. 1–23, 2016.
50. A. Radziwon, A. Bilberg, M. Bogers, E. S. Madsen, "The smart factory: Exploring adaptive and flexible manufacturing solutions," *Procedia Eng.*, vol. 69, pp. 1184–1190, 2014.
51. A. Mussomeli, D. Gish, S. Laaper, "The rise of the digital supply chain," *Deloitte*, vol. 45, no. 3, pp. 20–21, 2015.
52. S. Jha, "Hero moto corp: How Vijay Sethi is driving the digital twin project at Hero Moto Corp, IT news, ET CIO," 2017. [Online]. Available: https://cio.economictimes.indiatimes.com/news/strategy-and-management/how-vijay-sethi-is-driving-the-digital-twin-project-at-hero-moto-corp/57625617. [Accessed: 29-Nov-2018].

53. S. Goldberg, "The promise & challenges of digital twin," *HarperDB*, 2018. [Online]. Available: https://www.harperdb.io/blog/the-promise-challenges-of-digital-twin. [Accessed: 19-Nov-2018].
54. David Schahinian, "IoT forecast: Digital twins to be combined with blockchain - digital twin - HANNOVER MESSE," 2018. [Online]. Available: http://www.hannovermesse.de/en/news/iot-forecast-digital-twins-to-be-combined-with-blockchain-88960.xhtml. [Accessed: 06-Sep-2018].
55. S. Goldberg, "The promise & challenges of digital twin," 2018. [Online]. Available: https://www.harperdb.io/blog/the-promise-challenges-of-digital-twin. [Accessed: 30-Nov-2018].
56. S. Ferguson, E. Bennett, A. Ivashchenko, "Digital twin tackles design challenges," *World Pumps*, vol. 2017, no. 4, pp. 26–28, 2017.
57. S. Haag, R. Anderl, "Digital twin – Proof of concept," 2018.
58. R. Adams, G. Parry, P. Godsiff, P. Ward, "The future of money and further applications of the blockchain," *Strateg. Chang.*, vol. 26, no. 5, pp. 417–422, 2017.
59. A. Shakir, Zeeshan Muhammad;Aijaz, "IoT, robotics and blockchain: Towards the rise of a human independent ecosystem | IEEE communications society," 2018. [Online]. Available: https://www.comsoc.org/publications/ctn/iot-robotics-and-blockchain-towards-rise-human-independent-ecosystem. [Accessed: 04-Dec-2018].
60. J. K. Hodgins, "Animating human motion," *Sci. Am.* vol. 278. Scientific American, a division of Nature America, Inc., pp. 64–69, 1998.
61. S. Tyagi, A. Agarwal, P. Maheshwari, "A conceptual framework for IoT-based healthcare system using cloud computing," In *2016 6th International Conference - Cloud System and Big Data Engineering (Confluence)*, pp. 503–507, 2016.
62. F. Liu, A. Wollstein, P. G. Hysi, G. A. Ankra-Badu, T. D. Spector, D. Park, G. Zhu, M. Larsson, D. L. Duffy, G. W. Montgomery, D. A. Mackey, S. Walsh, O. Lao, A. Hofman, F. Rivadeneira, J. R. Vingerling, A. G. Uitterlinden, N. G. Martin, C. J. Hammond, M. Kayser, "Digital quantification of human eye color highlights genetic association of three new loci," *PLoS Genet.*, vol. 6, no. 5, p. e1000934, 2010.
63. J. Davis, H. Bracha, "Prenatal growth markers in schizophrenia: A monozygotic co-twin control study," 1996.
64. D. Baars, "Towards self-sovereign identity using blockchain technology," University of Twente, p. 90, 2016.
65. V. Gohil, "Blockchain's potential in India • indiaChains," 2018. [Online]. Available: https://indiachains.com/blockchains-potential-in-india/. [Accessed: 29-Nov-2018].
66. S. Haridas, "This Indian city is embracing blockchain technology -- here's why." [Online]. Available: https://www.forbes.com/sites/outofasia/2018/03/05/this-indian-city-is-embracing-blockchain-technology-heres-why/#337fcfb88f56. [Accessed: 29-Nov-2018].
67. e-Estonia, "Frequently asked questions: Estonian blockchain technology," 2017.
68. "Government of Canada ' s Innovation supercluster initiative."
69. "Digital government 2020 prospects for Russia."
70. Y. Handoko, "Developing IT master plan for smart city in Indonesia," pp. 1–17.
71. C. Stolwijk, M. Punter, "Going digital: Field labs to accelerate the digitization of the dutch industry," 2018.

72. D. David, K. C. Lee, R. H. Deng, *Handbook of Blockchain, Digital Finance, and Inclusion. Volume 1, Cryptocurrency, FinTech, InsurTech, and Regulation*, ScienceDirect, 2017.
73. R. Woodhead, P. Stephenson, D. Morrey, "Digital construction: From point solutions to IoT ecosystem," *Autom. Constr.*, vol. 93, no. March, pp. 35–46, 2018.
74. P. Mamoshina, L. Ojomoko, Y. Yanovich, A. Ostrovski, A. Botezatu, P. Prikhodko, E. Izumchenko, A. Aliper, K. Romantsov, A. Zhebrak, I. O. Ogu, A. Zhavoronkov, "Converging blockchain and next-generation artificial intelligence technologies to decentralize and accelerate biomedical research and healthcare," *Oncotarget*, vol. 9, no. 5, pp. 5665–5690, 2015.
75. A. Volkenborn, A. Lea-Cox, W. Y. Tan, "Digital revolution: How digital technologies will transform E&P business models in Asia-Pacific," In *SPE/IATMI Asia Pacific Oil & Gas Conference and Exhibition*, 2017.
76. M. Chiang, T. Zhang, "Fog and IoT: An overview of research opportunities," *IEEE Internet Things J.*, vol. 3, no. 6, pp. 854–864, 2016.
77. J. Huang, "Building intelligence in digital transformation," *J. Integr. Des. Process Sci.*, vol. 21, no. 4, pp. 1–4, 2018.

Chapter 11

Blockchain for Securing Internet of Things (IoT) Applications

Pramod Mathew Jacob and Prasanna Mani

Contents

11.1 Blockchain Concepts

Blockchain is an interesting technology which offers a secured mode for digital transactions. It acts like a 'distributed ledger' which records every transaction in a secure, auditable, efficient and transparent manner. This concept is new and has many applications and relevance in various business domains. Blockchain is simply a database system that holds a continuously growing distributed set of data records. Every transaction is digitally verified and signed to assure the authenticity. There is no master server which holds the entire chain. All the participating computers

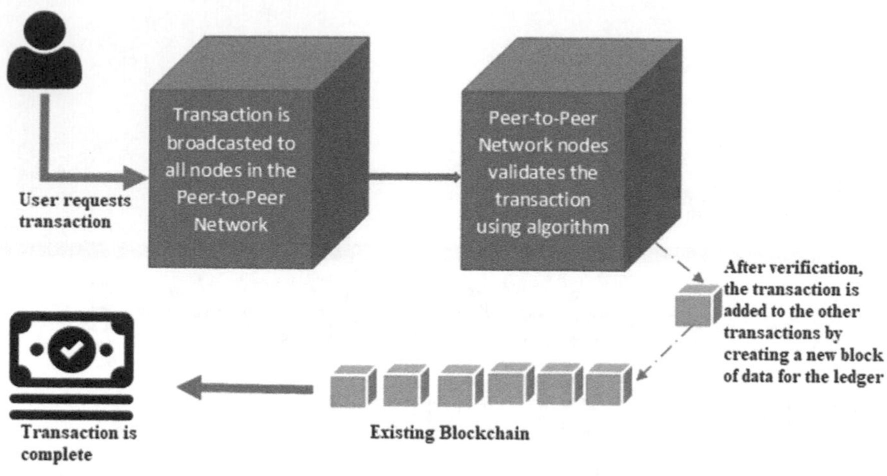

Figure 11.1 Illustration of working of blockchain technology.

(nodes) possess a copy of the transaction chain. Figure 11.1 illustrates the working of typical blockchain architecture [1].

A blockchain consist of two elements:

- **Transactions**: Any actions performed by the participants in the distributed system.
- **Blocks**: This element records all the transactions in sequential order and ensures that none of them have been tampered with. This is ensured by using a time stamp for all the transactions when and wherethey are added to the chain.

When a transaction edit request or a new transaction comes into blockchain, most of the nodes participating in the blockchain implementation run algorithms to verify and evaluate the history of every blockchain block that is considered. If the majority of the participating nodes feels the history and digital signature are valid, the new transaction block is accepted into the distributed ledger and a new block is appended to the transaction chain. If majority of the participating nodes doesn't feel the digital signature as authentic, then the change request or the addition request is denied and discarded. Thus this distributed consensus model permits blockchain to act as a distributed ledger which doesn't require some centralized authority to validate the records or transactions.

The three key properties of blockchain technology are

- Decentralization
- Immutability
- Transparency

In previous days there was a centralized system which may monitor and record all the transactions in a system. Any change can be initiated by the central coordinator. But the people working with a centralized system can tamper with the data of various transactions without the knowledge of other clients. This may lead to serious problems in case of a financial organization. Blockchain overcomes this demerit by providing a de-centralized system where the transaction chain is distributed among the participating clients or nodes. Whenever a node or a client tries to modify the data, it is intimated to all the other clients participating in that system. Thus it is impossible to tamper with the data without the consent of a majority of participating clients inside a blockchain. Thus decentralization became a key property of blockchain technology.

The property 'transparency' is a bit confusing for blockchain technology as it is considered as a secure system. Of course the system is secure and all the transactions and details of clients involved are stored in an encrypted form. But still if any clients tries to access or modify a transaction, all the participating clients will be alerted and thus it achieves transparency.

Immutability in blockchain is the property that ensures once a data is added to the system, it is impossible to tamper with the same. This is one of the unique properties of blockchain compared to other similar techniques like Bitcoin and centralized systems. Immutability is achieved in blockchain using some cryptographic hash function. Blockchain can be considered as a linked list which includes data and a hash pointer. The hash pointer points to its previous block and thus generates a chain of blocks. A hash pointer is similar to a pointer in a linked list, but instead of simply holding the address of the previous block it also holds the hash of the data inside the previous block existing in the chain.

The blockchain network is simply a collection of nodes which are interconnected. The blockchain is maintained by peer–peer network architecture. In a peer–peer model, there is no single centralized server. Every system participating in the network has equal priority. Every system can communicate with the others. The same system can work as both client as well as server in different instances. Thus there will be multiple distributed and decentralized servers. Though the system is using a peer–peer model, there will not be a single point of failure.

A node in the blockchain can be categorized as follows:

- **Light client**: A computer system which possess a shallow-copy of blockchain.
- **Full node**: A computer system possess the full-copy of blockchain.
- **Mining**: A computer system which verifies the transactions.

The various application domains of blockchain technology include

- Smart contracts
- Crowd funding
- Supply chain auditing

- Prediction markets
- File storage
- Internet of Things (IoT)
- Identity management
- Protection of intellectual property
- Anti-oney aundering (AML)
- Land title registration
- Stock market

This chapter further focuses on the application and scope of blockchain technology in Internet of hings (IoT)-based systems.

11.1.1 Internet of Things

The Internet has made a dramatic change in the field of information technology, making communication easier. As the world is using intelligent and smarter devices, technical experts derived the concept of 'Internet of Things'. The Internet of Things (IoT) is a networked class of devices, sensors and actuators deployed in distinct locations [2]. The connection between various components can be wired or wireless. Each device in the network should have a unique address. IPv6 protocol is used for the same as it can address up to millions of distinct devices. The typical IoT architecture is illustrated in Figure 11.2.

It is basically an inter-networked set of devices and embedded physical components. The physical system may include a microprocessor or a microcontroller.

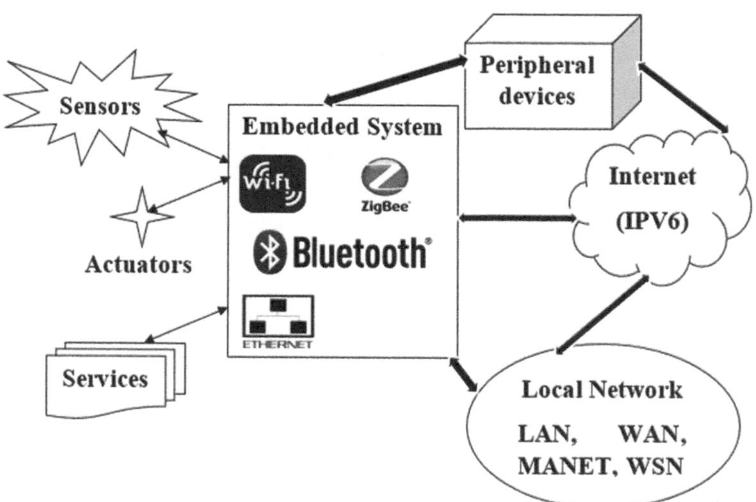

Figure 11.2 Typical architecture of IoT.

Arduino, Intel Galeleo and Raspberry PI boards are examples of the same [3]. Different sorts of sensors are deployed to collect real-time data. These fetched data are fed to the central coordinator device which processes the data accordingly and initiates suitable action using the connected actuators.

IoT uses both hardware and software. Apart from the hardware architectures, it uses a class of software architectural patterns. The various standardized software architectural patterns for IoT applications include lient–server, eer–eer, epresentational tate ransfer (REST) and ublish–ubscribe. The pattern selection criteria for different IoT applications is primely based on heterogeneity and security [4].

The Internet of hings has much relevance in the digital world where people like to control all objects and things remotely. Though the devices are heterogenous in nature, it is herculean task for the designer to model an IoT system for a particular domain. The various challenges in designing an IoT-based system are listed below:

- Compatibility and interoperability of heterogenous devices
- Lack of standardization in device identification and authentication
- Difficulty in integrating IoT applications with IoT platforms
- Difficulty in handling the unstructured, unformatted data
- Ensuring reliable connectivity between devices
- Information security and privacy concerns

One of the critical challenges in IoT-based systems is to ensureprivacy and data security. Many IoT applications like patient health monitoring, structure health monitoring (buildings, dams, etc.), weather forecasting, manufacturing and power plants deal with highly sensitive data. It is a risky job for the IoT developer to ensure privacy as well as security for these collected data. Blockchain can play a great role in this scenario. In the next section of this chapter, we discuss the integration of lockchain technology in IoT applications.

11.2 Integrating Blockchain in IoT

The Internet of Things (IoT) is transforming and efficiently optimizing the manual processes to obtain huge volumes of data collected from various real-time systems. These collected data are processed accordingly and the required information is extracted to derive conclusions. This model is used in weather forecasting, stock market prediction, smart farming, patient health monitoring, etc. The concept of cloud computing provides various functionalities to the IoT systems like data analysis and data processing. This unprecedented development in the IoT has paved the way for new mechanisms to access and share information. But due to the transparent nature of IoT systems, end users lack the confidence to share sensitive information through IoT systems. Centralized architecture is used in most of the IoT applications where the network participants don't have a clear vision of the shared data through the network. The shared information may look like a black box, and the users don't know the authenticity and source of data. The need for blockchain in IoT is discussed in the next section.

Due to the distributed nature of IoT networks, every node is a possible point of failure which may be exploited by cyber attackers (e.g. distributed-denial-of-service attacks). An integrated class of nodes with multiple infected devices working simultaneously may lead to system collapse. Another key concern is the presence of a central cloud service provider in an IoT environment. Any failure to this central node may lead to vulnerability which should be addressed. One of the most critical issues is data authentication and confidentiality. A lack of data security in IoT devices can be exploited and may be used in an inappropriate manner. Due to intervention of modern business models where the system can share or exchange data/resources autonomously, the need for data security is critical.

Another critical challenge in IoT is data integrity which has found some applications in the area of decision support systems (DSS). The collected data from the sensors can be used for generating timely instructions or decisions. Thus it is mandatory to protect the system from injection attacks where the attackers inject false measures or values into the system which may seriously affect the accurate decision making. Availability is critical for application domains, manufacturing plants, automated vehicular networks and smart grids where the real-time data are continuously monitored. Loss of data during a particular interval may result in the entire system failure. The integration of a security measure to publicly verify the audit trail will be beneficial for these sorts of systems. This can be easily achieved by the integration of blockchain.

The integration of various technologies like IoT, loud computing and lockchain into a single system has proven to be incomparable as it ensures both performance and security. The concept of implementing blockchain in IoT systems is a revolutionary step as it provides trusted data-sharing services where the data are reliable and traceable [5]. The source of the data being generated can be traced out at any stage and at the same time the data remain immutable.

In domains like smart cities and AI-based smart cars, reliable data are to be shared for the inclusion of new nodes (participants) in the system, thereby enhancing the services. Thus the implementation of blockchain can complement Internet of hings (IoT)-based applications with increased reliability and enhanced security. Though the IoT functionalities can be improved with the aid of blockchain, there is still a great number of research constraints and issues that are to be resolved.

11.3 Secure IoT Applications Using Blockchain

The various application domains where blockchain technology and Internet of hings (IoT) are clubbed together are listed below.

- **Supply chain and logistics**: A supply chain network system involves different stakeholders like raw material supplier, brokers, retailers, etc. Also it involves multiple payment receipts and invoices. The duration of a supply chain may extend over months. Due to the presence of multiple stakeholders, a delay in delivering will be a serious challenge. Thus companies are using IoT-enabled vehicles to track the live location and shipment process. Though the current supply chain management system lacks transparency and data security, it is possible to incorporate blockchain to enhance the traceability and reliability of the network. The information collected through sensors is then stored in blockchain. Various IoT sensors like PIR motion sensors, GPS trackers, RFID chips and temperature sensors collect the information from logistic vehicles/logistics and provide accurate details about the status of shipments. Sensor information is then stored in the blockchain and all new actions are noted as transactions. Hence it is no longer possible for a stakeholder to tamper with or modify the data which may make the supply chain system transparent and trustable.
- **Smart home**: Most of the smart home applications like intrusion detection systems, authentic access to rooms, remote controlling of devices and systems require personal details like biometric recognition, facial recognition, voice recognition, etc. All these data stored in a typical centralized data store are vulnerable to security threats. This can be resolved by using the concept of blockchain.
- **Automotive industry**: The automotive industry started using the concept of IoT for smart car parking in parking slots using some sort of e-wallets or bit currency. The time at which the vehicle parked in a particular slot is automatically estimated and the approximated charge is deducted from the e-wallet remotely. The integration of blockchain technology during this process may enhance the trust of the end users.
- **Pharmaceutical industry**: The problem of counterfeit drugs in the pharmaceutical sector is drastically increasing. A pharmaceutical company is responsible for manufacturing, developing and distributing medicines across the globe. Thus tracking the complete shipment process of drugs is not an easy task. The traceable and transparency features of blockchain technology can be used to remotely monitor the shipment of medicines from origin to destination. The data stored in the distributed ledger are time stamped and recorded by various stakeholders.
- **Agriculture**: Here the farmers deploy various sensors in the farm fields. The data fetched by the sensors are monitored by farmers, buyers, etc. All the data are represented as a block and are distributed among the farmers, buyers and consumers. By monitoring the data, farmers can initiate suitable measures to enhance the yield whereas the suppliers and consumers can decide whether to buy that crop or not based on the data analytics.

Apart from these areas, integrated IoT–blockchain systems are used in the stock market, the land registering process, online vehicle tracking and management in toll booths etc. The various challenges involved in implementing these concepts are discussed in the next section.

11.4 Challenges in Integrating Blockchain and IoT

- **Resource constraints**: Most of the available IoT platforms have limited computational and communication resources. A blockchain system requires excessive memory and storage resources for efficient execution. A low-powered IoT device with limited memory cannot withstand a heavy-weight blockchain technology which requires memory in GBs.
- **Bandwidth requirements**: Blockchain platforms have to continuously interact with other participants in the consensus process. Due to the decentralized mode of the consensus process, platforms in the chain network may exchange information about the blockchain for validation and creation of new nodes. End devices in the IoT architecture usually possess limited bandwidth. Thus this sort of processing is not easy in the end devices layer which may make the blockchain implementation difficult.
- **Security**: Though blockchain deals with decentralized architecture, all the devices in an IoT system may communicate and coordinate through a predefined protocol. Thus it is important for the IoT device to continuously participate in the blockchain which may make the device vulnerable to security issues and threats.
- **Latency demands**: IoT applications mainly include a set of data producers and consumers. In some instances the data consumer may initiate some actions. The introduction of blockchain technology may limit this freedom for the data consumer to initiate such actions though it may be considered as some sort of tampering in a blockchain system. Thus it cannot be applied in time-sensitive IoT applications.

11.5 Advantages of Using Blockchain in IoT

- Trust between smart devices and third parties is enhanced and assured.
- More economical for industrial and business applications.
- Transaction settlement time frame can be reduced.
- Improved data consistency.
- Enhanced cyber security.

11.6 Related Works in IoT and Blockchain

There have been many researchers who exploited the benefits of integrating blockchain in IoT. Securing the data exchanged between IoT devices is a critical challenge for all IoT service providers. Though various security measures exist, IoT requires a light-weight security model to ensure data integrity and security.

Kim et al. [6] proposed a taxonomy for securing IoT devices used in home and business applications. They encrypted the data shared in the distributed IoT architecture and a smart contract is used to ensure the data integrity. They validated their system with a home automation system. Their experimental results prove that various security threats like man-in-the-middle attacks, data stealing, etc., can be avoided with the help of blockchain.

Fakhri et al. proposed [7] a comparison model of a smart refrigerator system with and without blockchain technology. They initiated explicit sniffing attacks to prove the validity of their model. The experiment results claim that blockchain has the upper hand over traditional security measures in an IoT system. They have observed the avalanche effect of encryption algorithms and the hash functions used. MQTT is used as the software pattern for application without IoT.

Oscar Novo [8] proposed a detailed clear-cut implementation of incorporating blockchain in IoT. This lightweight, transparent and scalable model introduces a new access control policy among the stakeholders using the benefits of blockchain. A node called the management hub is introduced in IoT to store the various distributed smart contract information. They have implemented their model with the help of Ethereum which is one of the most popular blockchain platforms.

Pin et al [9] proposed an publish–subscribe-based IoT model over blockchains. This model mainly focuses on centralized IoT systems where all the data are stored at a single point. The failure of this node may lead to the failure of the entire system. The data integrity of this sort of system can be ensured by using blockchain technology. They have implemented a light-weight, primitive key-based algorithm for ensuring the data security. They have validated their model with the help of the Ethereum platform.

Viriyasitavat et al. [10] proposed a blockchain-based service handling operations in Internet of Things. Their model claims that blockchain can be used for achieving interoperability of various services. They integrated together service-oriented architecture (SoA), blockchain technology (BCT) and various key performance indicators (KPI) which resolves both trust issues and interoperability challenges in the IoT system.

Doku et al. [11] proposed Lightchain which is a dedicated blockchain architecture for IoT. A proof of work (PoW) mechanism was initially used for verifying transactions. But the computational tasks and effort required to resolve a PoW puzzle are enormously high which is not acceptable in a light-weight architecture like IoT. The PoW puzzle-solving efforts are distributed among various nodes in the IoT system. Thus the overhead of every single node can be drastically reduced which will improve the overall system performance and security.

Pan et al [12] proposed Edgechain which is an edge computing-based IoT architecture which incorporates the blockchain technology. The central node of the IoT architecture is excluded from the computational overhead of blockchain technology. All these operations are performed in the edge-based cloud pool which makes the architecture light-weight. Thus it can ensure features like data security, integrity, scalability, interoperability and enhanced performance aspects.

11.7 Summary

Blockchain is a promising technology for ensuring the data security and trustworthiness of the end user. Though IoT is incorporated in almost all aspects of human life, the security of personal data is the primary concern of every end user. Though IoT uses light-weight architecture, it is not easy to use strong security algorithms to prevent data stealing. In this scenario blockchain came to rescue IoT applications by providing a light-weight decentralized, distributed architecture to secure the data. The implementation of Edgechain and Lightchain proves that blockchain and IoT can go a long way in the future years of computer technology.

References

1. Blockgeeks. [Online]. 2019, May. https://blockgeeks.com/guides/what-is-blockchain-technology/.
2. Vijay Madisetti, Arshdeep Bahga, *Internet of Things: A Hands on Approach*. Universities Press, First edition, 2015.
3. Prasanna Mani, Pramod Mathew Jacob, "A Reference Model for Testing Internet of Things Based Applications," *Journal of Engineering, Science and Technology (JESTEC*, vol. 13, no. 8, pp. 2504–2519, 2018.
4. Prasanna Mani, Pramod Mathew Jacob, "Software Architecture Pattern Selection Model for Internet of Things Based Systems," *IET Software*, vol. 12, no. 5, pp. 390–396, October 2018.
5. Cristian Martín, Jaime Chen, Enrique Soler, Manuel Díaz, Ana Reyna, "On Blockchain and Its Integration with IoT. Challenges and Opportunities," *Future Generation Computer Systems*, vol. 88, pp. 173–190, November 2018.
6. M. Singh, A. Singh, S. Kim, "Blockchain: A Game Changer for Securing IoT Data," In *2018 IEEE 4th World Forum on Internet of Things (WF-IoT)*, Singapore, pp. 51–55, 2018.
7. D. Fakhri, K. Mutijarsa, "Secure IoT Communication Using Blockchain Technology," In *2018 International Symposium on Electronics and Smart Devices (ISESD)*, Bandung, pp. 1–6, 2018.
8. O. Novo, "Blockchain Meets IoT: An Architecture for Scalable Access Management in IoT," *IEEE Internet of Things Journal*, vol. 5, no. 2, pp. 1184–1195, April 2018.
9. L. Wang, H. Zhu, W. Deng, L. Gu P. Lv, "An IoT-Oriented Privacy-Preserving Publish/Subscribe Model Over Blockchains," *IEEE Access*, vol. 7, pp. 41309–41314, 2019.
10. L. Da Xu, Z. Bi, A. Sapsomboon, W. Viriyasitavat, "New Blockchain-Based Architecture for Service Interoperations in Internet of Things," *IEEE Transactions on Computational Social Systems*, vol. 6, no. 4, pp. 739–748, August 2019.
11. D. B. Rawat, M. Garuba, L. Njilla, R. Doku, "LightChain: On the Lightweight Blockchain for the Internet-of-Things," In *2019 IEEE International Conference on Smart Computing (SMARTCOMP)*, Washington, DC, USA, pp. 444–448, 2019.
12. J. Wang, A. Hester, I. Alqerm, Y. Liu, Y. Zhao, J. Pan, "EdgeChain: An Edge-IoT Framework and Prototype Based on Blockchain and Smart Contracts," *IEEE Internet of Things Journal*, vol. 6, no. 3, pp. 4719–4732, June 2019.

Chapter 12

Bitcoins and Crimes

M. Vivek Anand, T. Poongodi, and Kavita Saini

Contents

12.1 Introduction

The Internet is a data warehouse where the content is huge and browsers provide the service of retrieving the data from the Internet. Although the data on the Internet are huge, the retrieval of data by most of the browsers is minimized. Only 5% of the data is retrieved from the Internet that is indexed. Even though search engines like Google have an enormous amount of data, all the data cannot be retrieved because they are not indexed. The information which is retrieved from the Internet is indexed and it is called the surface web. The remaining 95% [7] of the web is called the deep web. Four percent of the highly confidential information on the deep web is called the dark web. The deep web contains confidential data like bank details, Facebook private details, etc., that should not be visible to all the people in the world. The criminals are accessing dark web for criminal activities with bitcoins as a transaction currency. I2P (Invisible Internet Project), Freenet and Tor browser provide access to the dark web that cannot be accessed by the normal browser. I2P, Freenet and Tor browser is not created for malicious access. It was created with the intention of providing anonymity and protecting private data from the Internet. Anonymity provides the protection of data from the Internet by hiding the user's identity on the Internet, but this is leading to criminal activities. The Tor browser is used for anonymity, and it is also called onion routing.

12.1.1 Background

The Naval Research Laboratory in the US developed an onion routing technique in 1998. On September 20, 2002, the Tor browser was released publicly for use on the Internet. With a normal browser like Google, the requested data are sent to the Internet service provider and to the Domain Name System where it will be verified with the IP address. Each website address on the Internet has an IP address to access the website. The Domain Name System provides the service of changing the website address to an IP address on request and changing the IP address to the website address in response.

The Tor browser provides anonymity [19] because search details are not known to the Internet service provider. The Internet service provider knows only which browser is accessing now but does not know what content is accessed. The Tor browser transfers the data, though nodes, which are already connected in the Tor network. The relaying of data in the network node passes through different places because the nodes are in different places all over the world. The Tor browser provides encrypted data by https add-on service because http is not encrypted. Even though onion routing provides anonymity and protects private data, criminal activities are increased because of the anonymity. Criminals use the Tor browser to do illegal activities like selling guns, drugs, etc. Hitman Network is a website that is used for criminal activities such as selling drugs with the exchange of money. These criminal activities with exchange of money leads to problems for the criminals when it is transferred through a bank. Criminals carry out the transaction with cryptocurrency to maintain anonymity. Cryptocurrency has an encryption mechanism to protect money from hacking. Bitcoin is the first cryptocurrency that works with the blockchain network. The Bitcoin blockchain is a distributed ledger that has all the transaction details. The cryptocurrency network carries out the transaction without a third party, such as banks. Bitcoin was created by Satoshi Nakamoto and he is the first man entered into the blockchain network. In banking, the transaction bank will act as a trusted third party, and it governs all the transactions over the network. The dependency on a trusted third party is essential for bank transactions. If a bank robbery or a hacking of bank database happens. That will lead to an uncontrollable situation. To avoid the third-party trust cryptocurrencies came into existence.

12.1.2 Introduction of Bitcoin

A cryptocurrency, as described by Satoshi Nakamoto in 2008 [1] and introduced as open-source software in 2009, is a collection of technologies that form digital or cryptocurrency. The units of this cryptocurrency are called Bitcoin. Bitcoin works with the concept of blockchain to avoid the double-spending problem in transactions [2]. It is used to store and transmit value among all the participants in the Bitcoin network. It is a peer-to-peer technology which is not governed by any central authority or bank. Unlike traditional currencies, Bitcoins are entirely secure and virtual. There are no physical coins available, rather it is an entirely virtual currency. The system is run by the Bitcoin protocol, and it is based on mathematics, unlike conventional currencies that are based on a fixed quantity or fiat currencies.

12.1.3 Bitcoin Features

Bitcoin has several features that set it apart from fiat currencies:

- Released by Satoshi Nakamoto in 2008.
- It is a decentralized and distributed secure digital currency.
- It is very easy to set up and fast.
- It is anonymous and completely transparent.
- Transactions are irreversible.
- A distributed transaction log is maintained for all participants.

The basis of the Bitcoin protocol is a peer-to-peer system (Figure 12.1) which means that there is no need for a third party. Therefore, it is not controlled by a central authority but rather is created by a community of people that anyone can join.

The Bitcoin protocol stores details of every single transaction that occurred in the network in a huge version of the general ledger (blockchain). Bitcoins are stored in a wallet with digital credentials for your Bitcoin holdings, allowing you to access them. The wallet uses public-key cryptography, in which two keys, one public and

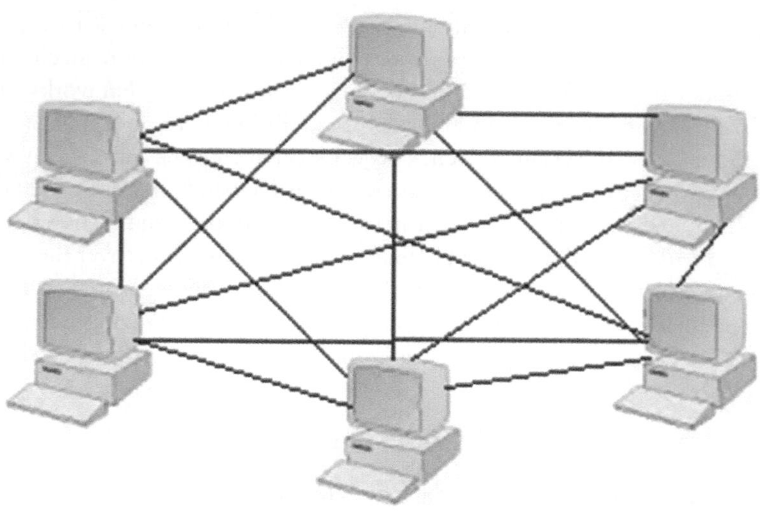

Figure 12.1 Peer-to-peer network.

one private, are generated. The public key can be thought of as an account number or name; the private keyis a ownership credentials.

Bitcoin is transferred to the next owner when the next owner gives a public key and the previous owner uses his private key to publish a record into system announcing that the ownership has changed to the new public key. Unlike transactions through the bank, Bitcoin doesn't charge fees for any transfers at the national or international level [9]. Bitcoin protects against double-spending [11] by verifying each transaction added to the blockchain to ensure that the inputs for the transaction had not previously already been spent.

12.1.4 Blockchain and Bitcoin

The blockchain is the technology behind Bitcoin. Bitcoin is the digital token, and blockchain is the ledger that keeps track of who owns the digital tokens. You can't have Bitcoin without blockchain, but you can have a blockchain without Bitcoin. Other prominent cryptocurrencies are Ethereum [13, 14], Bitcoin Cash, Ripple and Litecoin.

12.1.5 Bitcoin Security

Figure 12.2 shows a representation of Bitcoin security. The biggest challenges of bitcoin [5] are to check authentication, integrity, availability and confidentiality. All these challenges are taken care of very smoothly.

- Authentication: Public key crypto: Digital signatures
- Integrity: Digital signatures and cryptographic hash

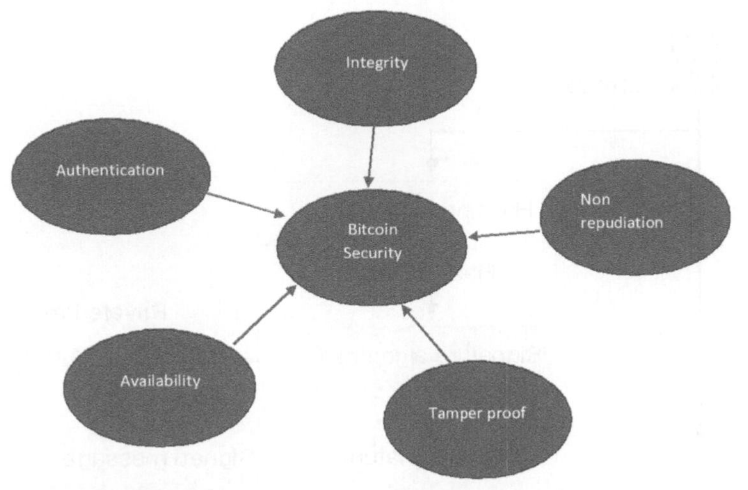

Figure 12.2 Bitcoin security.

- Availability: Broadcast messages to the P2P network, confidentiality: Pseudo anonymity

It is based on public key crypto (Figure 12.3): Encryption uses public key and private key. Public key crypto or digital signature are used to make sure that it is secure. First, create a message digest using a cryptographic hash and, then, encrypt the message digest with your private key.

12.1.6 Bitcoin Transaction

The most important part of the Bitcoin system is the transactions. Transactions are basically data structures that encode the transfer of value between participants evolved in the Bitcoin system [8]. It is ensured that transactions can be created and broadcasted on network. Once broadcasted on network they are added to the global ledger of transactions (blockchain) after validation.

12.1.7 Transaction Lifecycle

Various activities are involved in the transaction lifecycle (Figure 12.4), starting from the origin to recorded on the blockchain. The origin is basically the creation of the transaction. Once it is created it must be signed and authorized to spend funds referenced by the transaction. Once the transaction is authorized it is broadcasted to the network for validation. Finally, the transaction is verified by a mining node and included in a block of transactions that is recorded on the blockchain.

- Creation of transaction
- Broadcast on network

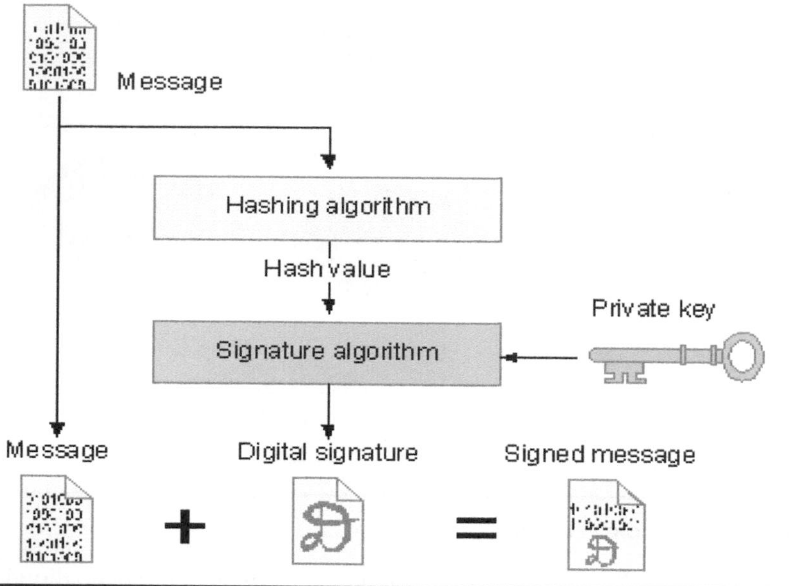

Figure 12.3 Public key crypto.

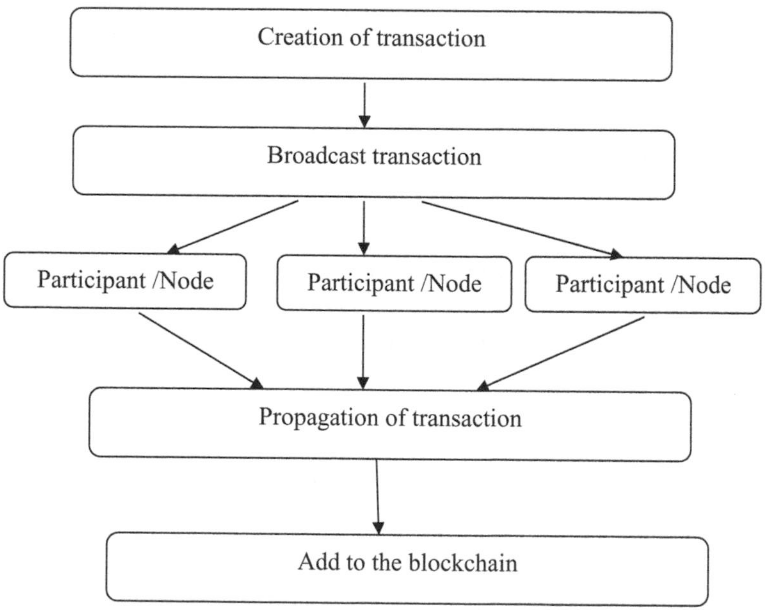

Figure 12.4 Transaction lifecycle.

- Verify transaction by many nodes
- Propagation
- Add to the blockchain

The **creation** of a transaction includes input, output, signature and amount.

Broadcast on network refers to sending the transaction to the neighboring node.

Verify transaction plays a very important role as many nodes or participants verify the transaction in this phase.

If the transaction is valid, it is added in blockchain and **propagated** to the entire network

Transactions in Bitcoin will be made by sending electronic payments. Transactions are the tiny building blocks of the blockchain system [Figure 12.5], and they consist of a sender address, recipient address and coins as in a normal credit card transaction. A Bitcoin transaction shifts the coin from one user wallet to another, and the coins are viewed as transactions, in particular, a chain of transactions. The destination address (Bitcoin address) is obtained by performing hashing operations using the user's public key. Every user in Bitcoin can have multiple addresses by generating several public keys; the addresses can be associated with users' wallets. The owned Bitcoins can be transmitted as digitally signed transactions by using the private key of the user. Moreover, it is highly recommended to use different Bitcoin address for each transaction.

The transactions in Bitcoin [Figure 12.6] will be verified for correctness, integrity and authenticity by the set of network nodes called "miners". The miners will

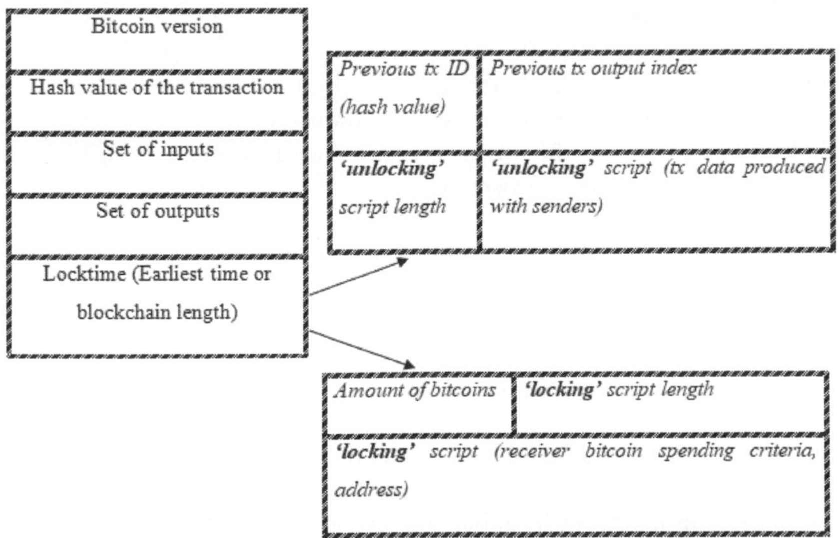

Figure 12.5 Fields in Bitcoin transaction.

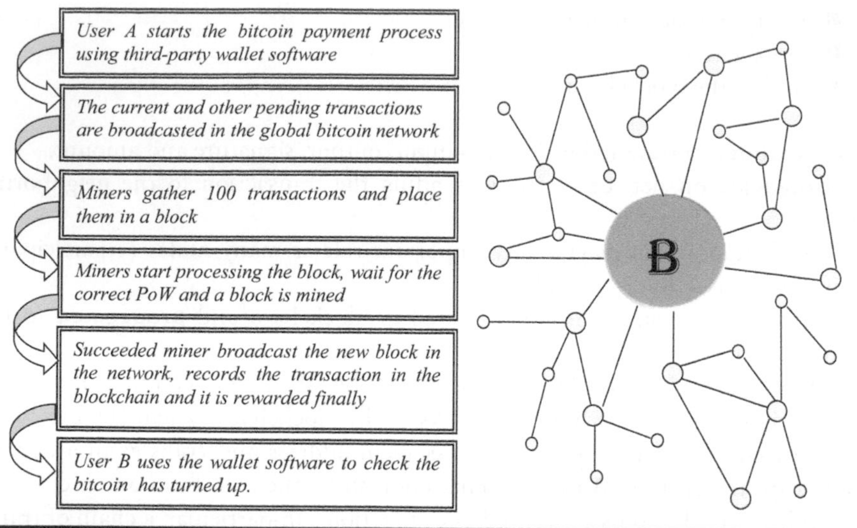
User A starts the bitcoin payment process using third-party wallet software
The current and other pending transactions are broadcasted in the global bitcoin network
Miners gather 100 transactions and place them in a block
Miners start processing the block, wait for the correct PoW and a block is mined
Succeeded miner broadcast the new block in the network, records the transaction in the blockchain and it is rewarded finally
User B uses the wallet software to check the bitcoin has turned up.
B

Figure 12.6 Bitcoin transaction.

collect some transactions as a single unit known as a "block" that are in the queue for processing. A block will be broadcasted in the entire network to claim the reward after completing the validation and mining process. Before updating the newly mined block in a public ledger, it will be verified by most of the miners in the network. The miners collect a reward once it is successfully added into blockchain. The significant characteristics of technical components that are essential in the Bitcoin system are presented.

The transaction input consists of:

1. Hash pointer (i.e., identifier that contains the output) to previous transaction that is taken as the input for the current transaction.
2. An index specifying the unsent previous transaction output (UTXO) that can be used in current transaction.
3. Unlocking script length.
4. Unlocking script (satisfies the condition associated with UTXO).

The transaction output comprises of:

1. Number of Bitcoins being transferred.
2. Locking script length.
3. Locking script (condition should be met before spending UTXO).

Every transaction input can be authorized using the corresponding user's public key, and the cryptographic signature is created using a private key. The input values from previous transactions which are listed in a single transaction can be added up and the sum is used by the output of the current transaction. In Bitcoin, the output of a previous transaction is exploited as the input of the current transaction; often the coin value may be higher than the user desires to pay. In this case, the sender creates a new Bitcoin address for getting back the difference amount. For example, user B has 100 coins from any of the previous transaction output, and he would like to transfer 10 coins to user A using the same output as an input in the current transaction. Particularly, user B has to generate a new transaction with an input (i.e., the output that user B has received the 10 coins) and two more outputs. In the output, one shows that 10 coins are being transferred to user A, and the other shows the transfer of remaining coins in any one of the wallets owned by user B.

Hence, the Bitcoin achieves two objectives:

1. It employs the idea of change.
2. The detail about the balance or unspent coins of a user can be identified by knowing the outputs from previous transactions.

An output in every transaction denotes the number of coins being transmitted along with the Bitcoin address of the new owner. Inputs and outputs in Bitcoin are handled using a script language to assert the Bitcoins. There are two main scripts in today's market:

1. Pay-to-PubKeyHash (P2PKH): In this script language, only one signature from the owner is required to authorize a payment.
2. Pay-to-ScriptHash (P2SH): This scheme uses multi-signature addresses; however it supports a variety of transaction types.

12.1.8 Overview of Bitcoin Architecture

Blockchain is an immutable database in which all Bitcoin transaction records are stored in chronological order. Tampering of data is not possible because of the hash technique used in each block creation. Computer systems in blockchain connected with each other adhered to the shared data and agreed to certain conditions imposed on it. In the beginning, Bitcoin was implemented as a large-scale blockchain. Nowadays, Bitcoin's blockchain is "simple" when compared to other blockchains. The ideas of different blockchain projects are growing rapidly. However, the Bitcoin architectural components are a digital signature, blockchain, distributed network and mining [12]. A digital signature is an asymmetric cryptographic technique created by the user's private key to assure the corresponding Bitcoin address. Bitcoin is also referred to as a cryptocurrency since the digital signature is a category of cryptographic techniques.

A blockchain is a decentralized, shared, distributed state machine, and all nodes in blockchain will hold their own copy independently. The current known "state" is dependent on each transaction processing. Bitcoin is a decentralized electronic payment system in which the nodes communicate using the P2P network. It uses a probabilistic distributed consensus protocol to achieve consensus among the communicating nodes. A centralized private ledger is maintained in a central bank to verify the process and record all transactions, whereas in Bitcoin, every user maintains their own copy of the ledger in blockchain. Vulnerabilities arise because of maintaining multiple copies of blockchain at many nodes in the network as it provides a consistent global view in the blockchain system. For example, user A could simultaneously create two different transactions to two different users, user B and user C, using the same set of coins. This malicious activity is referred to as double-spending [2]. In this instance, if both receivers are processing the transaction independently and the transaction verification process is successful, it leads to an inconsistent state. Bitcoin uses consensus protocol and proof of work (PoW) to satisfy the below-mentioned requirements:

1. The transaction verification process can be distributed among miners for ensuring the correctness of the transaction.
2. The successfully processed transaction should reach everyone in the network quickly to assure the consistent state of the blockchain.

Every distributed transaction process checks that the majority of miners [4] verify the authenticity of it before the transaction is added to the blockchain. If there is any update in the blockchain, the local copy which is maintained in all nodes will get updated; the correct state is attained by considering the majority of miners' agreement. Still, this system is vulnerable to Sybil attacks. In this type of attack, multiple virtual nodes can be created by the miner and those nodes will start sending false information in the network as positive votes for the faulty transactions to interrupt the election process. The countermeasure for Sybil attack in Bitcoin is to use PoW based on a consensus model; some computational tasks have to be accomplished by the miners to prove that they are real entities. PoW enforces a high-level computational cost for every transaction verification process, and the verification is based on the computing power of a miner. Faking computer resources is harder than performing a Sybil attack in the network.

A block is created by collecting pending transactions instead of mining individual transactions. A block is mined by computing the hash value with a varying nonce. A different nonce value is taken every time until the hash value becomes lower than or equal to the targeted value. The target is taken as a 256-bit number that is shared among all miners. Computing the desired hash value is extremely challenging. Bitcoin uses SHA-256 for calculating the hash value. Every time, different nonces (random values) are used for finding the required hash value until the solution is obtained. The correct hash value has been found by the miner for a block and the block is immediately broadcasted in the network with the computed hash value along with nonce. The remaining miners quickly verify the correctness of the received block by comparing the hash value with the target value. They will update the local blockchain by appending the newly mined block.

A block will be added successfully to the blockchain, once the majority of miners agree on the block as valid. The miner who got the solution for PoW [5] will be rewarded with a set of recently generated coins. Because of the absence of a central authority, the rewards do not reach anyone in the network. Rather, the rewards will be given for the block generation process, in which a coinbase transaction is inserted by the miner for the Bitcoin address, and it appears to be the first transaction in every block. Once the mined block is approved by the peers, then the newly inserted transaction becomes valid, and the miner obtains the awarded Bitcoins.

The Bitcoin network is usually not having transaction fee for their transactions, and it is only mentioned by the owners of a transaction and it also varies for every transaction. However, the transaction fees are increasing to some extent, discouraging the usage of Bitcoin. The security issues in Bitcoin that might occur if block rewards did not exist are investigated [17].

Blockchain is a public, linked-list based data structure which tracks the whole transaction history in the form of blocks. A Merkle tree structure [6] is followed for storing the transactions in every block, along with the secure time-stamp and hash value of the previous transaction.

12.1.8.1 Procedure to Add a New Block

1. Once the valid hash value for a block is determined by the miner, the block can be added to the user's local blockchain and the solution can be broadcasted.
2. If a solution is received for a valid block, the miners will immediately verify the validity, and if the solution is found to be correct the local copy is updated by the miner; otherwise, the block is discarded.

For mining, a dedicated application-specific integrated circuit is used by a single home miner in which a single block takes time to verify. Hence, for this reason, mining pools are introduced. In this, the set of miners can be associated together to mine a particular block beneath the control of the pool manager. Once the mining is successful, the manager issues the reward to the associated miners according to the number of resources spent by each miner.

12.2 Consensus Protocol

For assuring continuous service without any interruption, the fault-tolerant consensus protocol is essential to ensure that the participating nodes are agreed on the order. The set of rules mentioned in the consensus protocol should be followed by the miner to append a new block in the blockchain. Bitcoin obtains the distributed consensus based on PoW and the consensus algorithm. The major rules followed in this algorithm are as follows:

1. Rational input and output.
2. Unspent outputs can only be used in each transaction.
3. Spent inputs should have valid signatures.
4. No coinbase (created by the miner) transaction outputs were used within 100 blocks.
5. No transaction inputs spent within a lock time before the block is confirmed.

Hence, blockchain-based Bitcoin is considered as robust and secure due to the consensus model.

Micropayment channel networks have been introduced to address the scalability issue by maintaining the block size unchanged. In this, the payment channel is established among two parties, who can pay on behalf of others which are not recorded in the blockchain. This off-blockchain mode of payment assists in processing the payment faster and suggests a way to track the money transfer among two entities. Yet, these payment channel networks face the set of challenges regarding user privacy [18, 21], processing concurrent payments, and routing.

12.3 Peer-to-Peer Network

The Bitcoin system follows the communication structure as an unstructured peer-to-peer (P2P) network using the non-encrypted persistent TCP connections. In an unstructured P2P network, the peers are arranged randomly in a flat or hierarchical manner. Time-to-live (TTL) search, expanding rings, random walks are used to find the peers who have interesting data items. In common, the unstructured overlay network is a highly dynamic network topology in which the peers can join and leave the network frequently. This type of network is best suited for the Bitcoin system to disseminate the information as soon as possible to reach the consensus in the blockchain. Shadow event discrete simulator assists in simulating large-scale Bitcoin networks in a single machine.

12.4 Role of Bitcoin in Crimes

The pseudo-anonymity of blockchain networks allows criminals to carry out illegal activities with Bitcoins. Bitcoins provide a hidden identity and Bitcoin plays a vital role in crimes. Criminals make transactions with cryptocurrency because of its anonymity. Although blockchain provide secure transactions by its consensus algorithm, criminals do criminal activities through Bitcoin [3]. Criminals are use websites to sell drugs and guns through Bitcoin transactions. All kinds of criminal activities are performed with Bitcoin transactions. Most countries do not accept

Bitcoin transactions because of the criminal activities performed through Bitcoin transactions, and it will be an issue for the government to take action against the criminal activities.

12.4.1 Bitcoin Exchange

Bitcoin provides a space to change black money into white. Bitcoin exchangers are used to exchange money. Some of the popular Bitcoin exchangers are:

1. Binance
2. Bittrex
3. KuCoin
4. Huobi Pro
5. Bibox
6. Poloniex
7. Bitmex
8. GDAX
9. LocalBitcoins
10. Kraken
11. Bitstamp

Bitcoin exchange websites also available for changing money. A few websites are:

1. Cex.io
2. CoinMama
3. Wirex
4. Bitit

12.4.2 Ransomware

Computer bugs infect and lock up a computer, server or mobile device, and the attacker demands a ransom to restore control of the device. The WannaCry ransomware attack, in May 2017, was a worldwide cyber-attack by the WannaCry ransomware cryptoworm, which targeted computers running the Microsoft Windows operating system by encrypting data and demanding ransom payments in the Bitcoin cryptocurrency. These kinds of ransomware attacks are happening every year, for example:

- In 2015, 1000 attackers per day.
- In 2016, 4000 attackers per day (300 percentage increase when compared to 2015).
- In 2017, the Wannacry ransomware attack locked files on 300,000 corporate and government computers in 150 countries, demanding $300 in Bitcoins to unfreeze the data.

Anonymity in Bitcoins allows the user to hide their face, and it is not an easy task to find them. Digital currency like U.S. dollars can be easily tracked when we do transaction because it has a centralized authority like a bank, but Bitcoin, which works under the concept of blockchain, does not depend on any centralized authority such as a bank. Bitcoin provides a decentralized network, in which nobody can control their money. The user can control their own money, and it is not an easy task to track the transaction because the ledger has only the address as a public key like

1. 1BvBMSEYstWetqTFn5Au4m4GFg7xJaNVN2
2. 3J98t1WpEZ73CNmQviecrnyiWrnqRhWNLy

Denial-of-service attacks are possible in the Bitcoin network [15]. Tampering with a Bitcoin transaction also requires more than 90% of users to give approval. In real time, it is highly impossible. In Russia, a ransomware payment made to an account in Bulgaria was traced, but the account was split into 12 different accounts in different cities of Russia with the command and control located somewhere in Germany. As many Bitcoin addresses can be created as you wish because there is no single point of the address issuing authority and address generation for transaction will not be affected by any environmental issues.

12.4.3 Tax Evasion

Tax is very important to run the government with good economy in a safer zone. Tax evasion is happening nowadays all over the world but it can be identified by the government to take necessary action against the evasion. Bitcoin provides a way to evade tax because the money can be easily exchangeable by exchangers and it will become white money. Another issue is that governments' and tax agencies' guidelines are unclear. Even in the U.S. there is a lot of confusion around the fixing of taxes for Bitcoins. An investigation is required to check the Bitcoin network and their transaction to fix the tax for Bitcoins. An illegal transaction has to be checked at least for targeted users to find the illegal transaction but in bitcoin blockchain it is very difficult. If tracking of a suspect user is possible then the government authorities will have the power to handle the issues in tax evasion.

Bitcoin exchanges handle a large amount of money, that is, more than billions of dollars of transactions. If there is cooperation between the authorities and the exchangers there is no need to shut down the Bitcoin transaction. If it is possible to get all the information about Bitcoin's illegal transactions and the users' device addresses, a complete study is required to find out whether there are any changes in wallets. The existing information about the Bitcoin wallets [16] from its site is only the device compatibility issues that either on mobile or system. The measurable strategies necessary to catch artifacts from awed Bitcoin wallet (the place a user might sign in with the username and password).

In 2015 an estimated 2.8 million people in the U.S. owned cryptocurrency but only 807 people reported Bitcoins on their 2015 taxes. The IRS defines cryptocurrency as a property, not a currency, meaning gains and losses must be reported by taxpayers on Form 8949. Though the IRS has guidelines for the cryptocurrency, some may not know or understand the rules. Thirty-six percent of Bitcoin investors claimed they would knowingly not report capital gains or losses from cryptocurrency on their 2017 taxes. Only 807 people in the U.S. reported crypto for tax in 2015. That is a very low number, of the 2.8 million individuals in the U.S. who owned a blockchain account. Some of the reasons for tax evasion are:

- Unnamed virtual wallets create anonymity.
- The mix of public and private keys makes it difficult to track illegal transactions.
- The decentralized system allows secure cross-border payments.
- The darknet is a hidden part of the deep web where illegal activity thrives.

12.5 Dark Side of Bitcoin Crimes

12.5.1 Darknet

The Internet is a data warehouse where the content is huge, and browsers provide the service of retrieving the data from the Internet. Although the data on the Internet are huge, retrieving data from most of the browsers is minimized. Only 5% of the data is retrieved from the Internet that is indexed. Even though the search engines like Google have an enormous amount of data, all the data cannot be retrieved because they are not indexed. The information which is retrieved from the Internet is indexed, and it is called the surface web. The remaining 95% of the web is called the deep web. Four percent of the highly confidential information of the deep web is called the dark web. The deep web contains confidential data like bank details, Facebook private details, etc., that should not be visible to all the people in the world.

The criminals are accessing dark web for criminal activities with bitcoins as a transaction currency. I2P (Invisible internet protocol), Freenet and Tor browser provide access to the deep web that cannot be accessed by a normal browser. I2P, Freenet and Tor software is not created for malicious access. It was created with the intention of providing anonymity and protecting private data from the Internet. Anonymity protects data from the Internet by hiding the user's identity on the Internet, but this leads to criminal activities. The Tor browser is used for anonymity, and it is also called onion routing.

Figure 12.7 shows the flow of Bitcoins in the darknet market from 2011 to 2018.

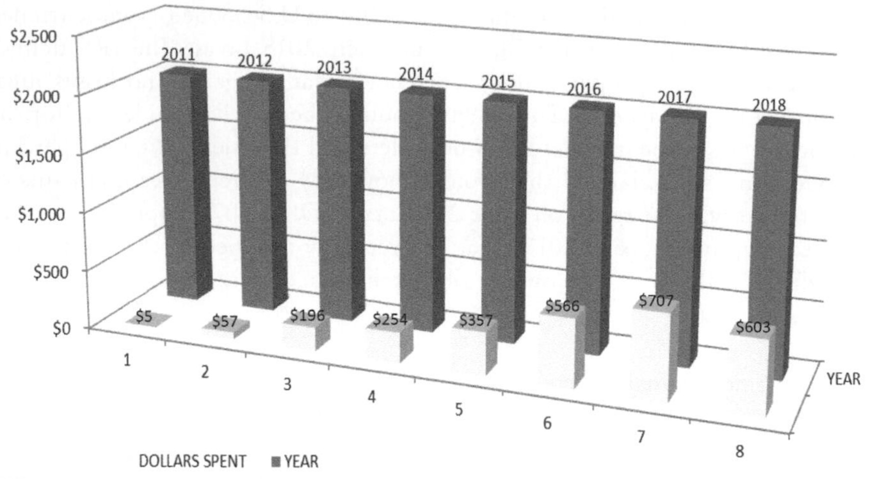

Figure 12.7 Flow of Bitcoins in darknet market in terms of dollars.

12.5.2 Tor Browser

Tor stands for The Onion Router, so-called because of the layered encryption process. Crypto-anarchism and onion routing are the two active terms linked to the underground web. Tor was originally founded by the U.S. Navy at the start of the millennium and is used by numerous agencies and others to transmit and receive sensitive information. Tor masks the user's identity and allows them to travel the surface web with total anonymity. Transferring money without leaving a trace is not always easy; however, the dark net's own currency Bitcoin provides the solution.

The Naval Research Laboratory in the U.S. developed the onion routing technique in 1998. The Tor browser was released publicly on September 20, 2002 for use on the Internet. In a normal browser like Google, the requested data are sent to the Internet service provider and to the Domain Name System, where it will be verified with the IP address. Each website address on the Internet has an IP address to access the Internet. The Domain Name System provides the service of changing the website address to an IP address on request and changing the IP address to the website address in response.

The Tor browser provides anonymity because searching details are not known to the Internet service provider. The Internet service provider knows only which browser is accessing now, but it does not know what content is accessed. The Tor browser transfers the data, through nodes, which are already connected in the Tor network. The relaying of data in the network node passes through different places because the nodes are in different places all over the world. The Tor browser provides encrypted data by https add-on because http is not encrypted. Even though onion routing provides anonymity and protects private data, criminal activities are increased because of the anonymity. Criminals use the Tor browser to carry out

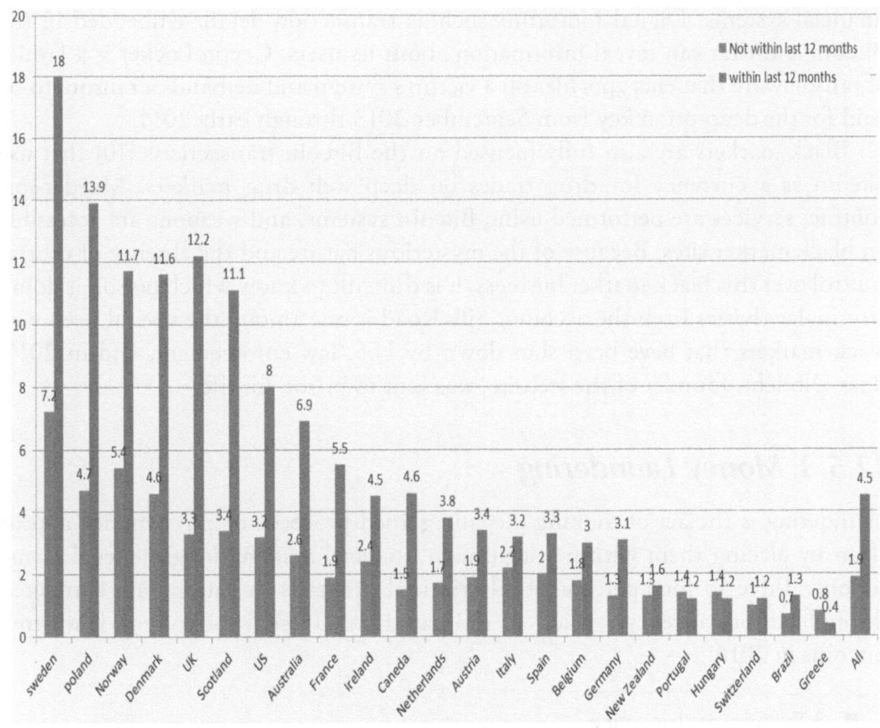

Figure 12.8 Drugs purchased report in darknet market.

illegal activities like selling guns, drugs, etc. Hitman Network is a website that provides criminal services by the exchange of money. This exchange of money for criminal activities leads to problems when it is transferred through a bank.

The dark web is the place that criminals use for illegal activities. In a survey, 9.3% of drug users bought drugs off the darknet in 2016, and 97.4% of illicit activity using Bitcoins originated from darknet marketplaces in 2013–2016. Criminals sell illegal drugs through sites like the first darknet market Silk Road, launched as the "eBay of drugs". There are:

- 13,000 drug listings.
- 1400 vendors.
- $1.2 billion of transactions, and the founder of Silk Road is Ross Ulbricht.

Figure 12.8 shows the number of people who bought drugs in the darknet market. The Tor hidden network and Bitcoin system [8] are undoubtedly useful in many aspects, playing vital roles in the cybercrime landscape. Distinguishing legitimate and illicit use of these services is a difficult task. This is how Bitcoin enables criminal enterprises to perform money laundering schemes compared to banks or financial systems. Digital footprints such as transaction details embedded in the Bitcoin and that can reveal information about its users. CryptoLocker is a family of ransomware that encrypts files on a victim's system and demands a ransom to be paid for the decryption key from September 2013 through early 2014.

Black markets are also fully focused on the Bitcoin transactions [10] that use bitcoin as a currency for drug trades on deep web drug markets. Murder-for-contract services are performed using Bitcoin systems, and weapons are accessible on black market sites. Because of the mysterious nature and the absence of central control over this black-market business, it is difficult to know which person is doing criminal activities from the account. Silk Road is one among the several deep web black markets that have been shut down by U.S. law enforcement, and in 2015, Ross Ulbricht, founder of the website, was sent to prison for life.

12.5.3 Money Laundering

Laundering is the act of making illegally gained proceeds or dirty money appear clean by placing them within a legitimate financial system alongside legal transactions. Three to four percent of all criminal proceeds in Europe are laundered through cryptocurrency, which is estimated at US$4–5 billion of laundered Bitcoins in 2016.

- 3.84% in multiservice
- 0.30% in crypto exchange
- 12.21% in gambling

Illegal online betting is much easier when money transfers can be made swiftly and untraceable.

- 0.05% in ATM
- 24.20% in the mixer
- 59.40% in the Bitcoin exchange
- Bitcoin laundering locations in 2016:
- 56.65% in Europe
- 36.44% in unknown jurisdictions
- 5.28% in North America
- 1.21% in Asia
- 0.35% in Oceania
- 0.07% in South America
- 0.00% in Africa

12.5.4 Scams and Fakes

Scams are created with Bitcoin advertisements such as offer price, bitcoin exchange, etc. Many users lose their currency and Bitcoins by the scams that are created by the

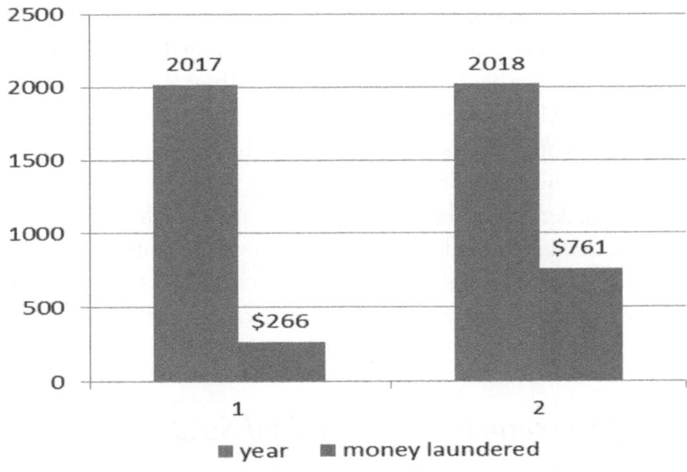

Figure 12.9 Variation in money laundering through cryptocurrencies.

illegal users. Increasingly, Bitcoin is becoming a staple utility among cybercriminals; two of the digital currency's main attractions are its provisions for pseudo-anonymity and its irreversible transaction protocol. These kinds of provisions engender the dichotomous incentives between legitimate users, who genuinely want to transfer money efficiently and securely, and cybercriminals who leverage these properties to commit irrevocable and supposedly untraceable transactions (Figure 12.9).

12.6 Open Challenges to Bitcoin Crimes

The anonymous and decentralized nature of cryptocurrencies has given opportunities for criminals to carry out illicit activities while evading prosecution. The international law enforcement community is responsible for cryptocurrency-related investigation cases and they receive the cases on a monthly basis from INTERPOL. Cryptocurrencies have been used extensively in darknet markets for receiving payments for illicit services such as distributed denial of service, malware binaries, botnets and the purchase of illegal products including weapons, drugs and falsified or stolen documents. Darknet markets such as Silkroad, AlphaBay and Hansa were reaping large profits, reaching $3,000,000 between September 2015 and December 2016. These sites facilitate the trade of illicit products like stolen antiquities, drugs and firearms, remitting money to areas that are under high financial scrutiny or embargo, and publicly crowd-fund their operations.

Money laundering is also a problem in cryptocurrencies, and it is a major challenge for law enforcement to solve the issues. A significant number of criminals are advertising cryptocurrency exchanges or initial coin offering with the goal of money laundering to get illicit profits. Bitcoin exchanges like OKCoin with hundreds of thousands of U.S. dollars laundered as well as the case of BitInstant, in which more than $1,000,000 was laundered for Silk Road customers. Cryptocurrencies have advanced the operations of different malware families such as ransomware, with CryptoLocker and CryptoWall receiving 133,045.9961 BTC and 87,897.8510 BTC, respectively; crypto jacking, with JenkinsMiner earning its operator over $3,000,000 worth of Monero; and crypto-stealing Trojans, such as CryptoShuffler, which stole hundreds of thousands of U.S. dollars by targeting the contents of volatile memory, that is, the clipboard.

The biggest challenge for law enforcement agencies here is the immutable nature of the blockchain that disallows the removal of embedded illicit content. Cryptocurrencies are used for sponsoring nation-state attacks, as a number of countries around the world are highly affected by the existence of contemporary hybrid-war strategies.

12.6.1 Law Enforcement and Criminal Strategies

Bitcoins are associated with a plethora of crime types such as narcotics, firearms, money laundering, terrorism and child exploitation; the international law enforcement community, including INTERPOL, has begun to focus on mastering the blockchain. In doing so, a significant amount of resources has been allocated for the exploration of the use of Bitcoins by criminals as well as the development of proprietary analytical tools for tracing Bitcoin transactions. In policing, two different schools of thought exist, with the first perceiving Bitcoins as a threat and the second viewing them as an investigational opportunity.

The first group considers Bitcoins a disruptive solution enabling criminals to facilitate their illegal activities in the absence of policing, and hence calls for its prohibition. The law enforcement community views cryptocurrencies as an investigation opportunity where criminally associated information is now publicly and permanently indexed in the blockchain to be analyzed for the extraction of valuable forensic data that can lead to attribution and prosecution.

Nowadays there has been a significant effort from both the industry and various law enforcement agencies, including INTERPOL, to develop forensic development tools and methodologies for the analysis of various cryptocurrencies such as Bitcoins. The vast focus on the analysis of Bitcoin transactions can be attributed to the substantial number of criminal cases affiliated with it, despite the existence of more anonymous cryptocurrencies. The market value of Bitcoin and its wide adoption by markets have played a catalytic role in boosting the magnitude of the criminal cases and trends associated with it. Despite the wide adoption of Bitcoins by criminals, the recent success stories of contemporary analytical tools that allowed police investigators to partially de-anonymize the Bitcoin network and reveal the identities of criminals have caused a shift in the use of cryptocurrencies. An increasing number of criminals are using Bitcoin only as an entry and exit point.

Law enforcement considers the cryptocurrencies highly disruptive due to their enhanced anonymity, which makes them an effective weapon for criminals. Dash and Zcash enable users to keep their activity history and balances private, which ultimately restricts law enforcement investigators from identifying and tracking suspicious transactions. Similarly, Monero uses ring signatures, ring confidential transactions and stealth addresses to obfuscate the origins, amounts and destinations of transactions.

Verge is another anonymous cryptocurrency that leverages the wraith protocol to enable its users to switch between public and private ledgers. When the wraith protocol is turned on, the transaction data are hidden. Finally, Namecoin does not share the same strong anonymity characteristics or goals as the cryptocurrencies above but is still identified as a potential threat to police due to its feature that allows criminals to anonymously register illegal websites without providing any personal information, thus preventing investigators from identifying the administrators behind these pages. As an extra layer of protection, many criminals use crypto-mixing/tumbling services or decentralized P2P exchange markets to "clean" their "tainted" coins, making it increasingly difficult for police investigators to follow their transactions.

To counteract the illicit use of cryptocurrencies, law enforcement is now focusing on the development of advanced solutions for tracing criminally linked transactions. In particular, police agencies work toward developing forensic tools for the analysis of various computing. It is imperative for law enforcement agencies to co-evolve with the current state of the art and identify and thwart online criminal activities that are linked to cryptocurrencies. www.computer.org/security 93 devices for the identification of cryptocurrency-related artefacts, such as wallets, and cryptocurrency hashes; fingerprinting tools for identifying the use of mixers/ tumblers in cryptocurrency transactions; clustering solutions for the aggregation of the addresses belonging to the same criminal actors for better attribution; and cross-ledger tracing tools to support the association of suspicious transactions in different blockchains.

In addition to the current work by law enforcement on a national level, INTERPOL acts as an information hub on the international level by bringing police investigators from various nations, researchers and blockchain developers together to share best investigation practices and forensic tools for cryptocurrencies.

INTERPOL works toward developing the investigation capabilities of its member countries by delivering advanced hands-on training on cryptocurrencies. INTERPOL strives for innovative solutions by seeking partnerships with the public and private sectors, including cyber security and cryptoanalytic companies.

INTERPOL held its first international Darknet and Cryptocurrencies Working Group in March 2018 to further stimulate discussions surrounding policing solutions for cryptocurrency-related crimes, identifying Altcoins and cross-ledger investigations as the biggest challenges for law enforcement.

12.7 Conclusion

Although numerous challenges handled by international law enforcement community and faces lots of investigating on cryptocurrencies, problems keep on increasing day by day. The blockchain is here to stay due to the wide use of cryptocurrencies by investors, adopters and pioneers. While it is anticipated that a large number of its characteristics will significantly adapt in the near future to overcome critical technological shortcomings such as its size and scalability, its use for illicit activities will only continue to grow. It is imperative for law enforcement agencies to co-evolve with the current state of the art and identify and thwart online criminal activities that are linked to cryptocurrencies. For better efficiency [20] in combatting cryptocurrency-related crime, an implementation of an international understanding and a legal framework to regulate it should be considered; this will enable law enforcement to access information for criminally linked transactions and urge cryptomarkets and exchanges to enforce strong KYC policies. The solution is required to access Bitcoin network without the criminal activities for smooth functioning of the blockchain with cryptocurrency in the future.

References

1. S. Nakamoto, "Bitcoin: A peer-to-peer electronic cash system," 2008, Available: http:// bitcoin.org/ bitcoin.pdf.
2. G. O. Karame, E. Androulaki, and S. Capkun, "Double-spending fast payments in bitcoin," In *Proceedings of the 2012 ACM Conference on Computer and Communications Security, ser. CCS '12*. New York, NY, USA: ACM, 2012, pp. 906–917.
3. A. Maria, Z. Aviv, and V. Laurent, "Hijacking bitcoin: Routing attacks on cryptocurrencies," In *Security and Privacy (SP), 2017 IEEE Symposium on. IEEE*, 2017.
4. I. Eyal and E. G. Sirer, "Majority is not enough: Bitcoin miningis vulnerable," In *Financial Cryptography and Data Security: 18th International Conference*. Berlin Heidelberg: Springer, 2014, pp. 436–454.
5. J. Bonneau, A. Miller, J. Clark, A. Narayanan, J. A. Kroll, and E. W. Felten, "Sok: Research perspectives and challenges for bitcoin and cryptocurrencies," In *2015 IEEE Symposium on Security and Privacy*, May 2015, pp. 104–121.
6. F. Tschorsch and B. Scheuermann, "Bitcoin and beyond: A technical survey on decentralized digital currencies," *IEEE Communications Surveys Tutorials*, vol. 18, no. 3, pp. 2084–2123, 2016.
7. W. F. Slater, "Bitcoin: A current look at the worlds most popular, enigmatic and controversial digital cryptocurrency," In *A Presentation for Forensecure 2014*, April 2014.
8. M. Kiran and M. Stannett, "Bitcoin risk analysis," Dec. 2014, Available: http:// www .nemode.ac.uk/ wp-content/ uploads/ 2015/ 02/2015-Bit-Coin-risk-analysis.pdf.
9. B. Masooda, S. Beth, and B. Jeremiah, "What motivates people to use bitcoin?" In *Social Informatics: 8th International Conference, SocInfo 2016*. Springer International Publishing, 2016, pp. 347–367.
10. K. Krombholz, A. Judmayer, M. Gusenbauer, and E. Weippl, "The other side of the coin: User experiences with bitcoin security and privacy," In *Financial Cryptography and Data Security: 20th International Conference, FC 2016, Christ Church, Barbados*. Berlin Heidelberg: Springer, 2017, pp. 555–580.
11. G. O. Karame, E. Androulaki, M. Roeschlin, A. Gervais, and S. Capkun, "Misbehavior in bitcoin: A study of double-spending and accountability," *ACM Transactions on Information and System Security*, vol. 18, no. 1, May 2015.

12. J. Heusser, "Sat solvingan alternative to brute force bitcoin mining," 2013, Available: https:// jheusser.github.io/ 2013/ 02/ 03/ satcoin.html.
13. G. Wood, "Ethereum: A secure decentralised generalised transaction-ledger," yellow paper, 2015.
14. A. Kosba, A. Miller, E. Shi, Z. Wen, and C. Papamanthou, "Hawk: The blockchain model of cryptography and privacy-preserving smart contracts," In *IEEE Symposium on Security and Privacy*, May 2016, pp. 839–858.
15. M. Vasek, M. Thornton, and T. Moore, "Empirical analysis of denial-of-service attacks in the bitcoin ecosystem," In *Financial Cryptography and Data Security: FC 2014 Workshops, BITCOIN and WAHC 2014*. Berlin Heidelberg: Springer, 2014, pp. 57–71.
16. "Biometric tech secures bitcoin wallet," no. 6, 2015.
17. M. Spagnuolo, F. Maggi, and S. Zanero, "Bitiodine: Extracting intelli-gence from the bitcoin network," In *Financial Cryptography and Data Security: 18th International Conference, FC 2014*. Berlin Heidelberg: Springer, 2014, pp. 457–468.
18. S. Goldfeder, H. A. Kalodner, D. Reisman, and A. Narayanan, "When the cookie meets the blockchain: Privacy risks of web payments via cryptocurrencies," *CoRR*, 2017.
19. A. Biryukov and I. Pustogarov, "Bitcoin over tor isn't a good idea," In *2015 IEEE Symposium on Security and Privacy*, May 2015, pp. 122–134.
20. S. Barber, X. Boyen, E. Shi, and E. Uzun, "Bitter to better — how to make bitcoin a better currency," In *Financial Cryptography and Data Security: 16th International Conference, FC 2012*. Berlin Heidelberg: Springer, 2012, pp. 399–414.
21. J. Herrera-Joancomart´ı and C. Perez´-Sola, "Privacy in bitcoin transac-tions: New challenges from blockchain scalability solutions," In *Model-ing Decisions for Artificial Intelligence: 13th International Conference, MDAI 2016*. Springer International Publishing, 2016, pp. 26–44.

Index